SpringerBriefs in Education

We are delighted to announce SpringerBriefs in Education, an innovative product type that combines elements of both journals and books. Briefs present concise summaries of cutting-edge research and practical applications in education. Featuring compact volumes of 50 to 125 pages, the SpringerBriefs in Education allow authors to present their ideas and readers to absorb them with a minimal time investment. Briefs are published as part of Springer's eBook Collection. In addition, Briefs are available for individual print and electronic purchase.

SpringerBriefs in Education cover a broad range of educational fields such as: Science Education, Higher Education, Educational Psychology, Assessment & Evaluation, Language Education, Mathematics Education, Educational Technology, Medical Education and Educational Policy.

SpringerBriefs typically offer an outlet for:

- An introduction to a (sub)field in education summarizing and giving an overview of theories, issues, core concepts and/or key literature in a particular field
- A timely report of state-of-the art analytical techniques and instruments in the field of educational research
- A presentation of core educational concepts
- An overview of a testing and evaluation method
- A snapshot of a hot or emerging topic or policy change
- An in-depth case study
- A literature review
- A report/review study of a survey
- An elaborated thesis

Both solicited and unsolicited manuscripts are considered for publication in the SpringerBriefs in Education series. Potential authors are warmly invited to complete and submit the Briefs Author Proposal form. All projects will be submitted to editorial review by editorial advisors.

SpringerBriefs are characterized by expedited production schedules with the aim for publication 8 to 12 weeks after acceptance and fast, global electronic dissemination through our online platform SpringerLink. The standard concise author contracts guarantee that:

- an individual ISBN is assigned to each manuscript
- each manuscript is copyrighted in the name of the author
- the author retains the right to post the pre-publication version on his/her website or that of his/her institution

More information about this series at http://www.springer.com/series/8914

Yuriko Yamamoto Baldin
Uldarico Malaspina
Editors

Mathematics Teacher Education in the Andean Region and Paraguay

A Comparative Analysis of Issues and Challenges

With Foreword by Jill Adler

International Commission on Mathematical Instruction
INSTITUTO DE INVESTIGACION SOBRE LA ENSEÑANZA DE LAS MATEMÁTICAS
PONTIFICIA UNIVERSIDAD CATÓLICA DEL PERÚ
Springer

Editors
Yuriko Yamamoto Baldin
Department of Mathematics
Universidade Federal de São Carlos
São Carlos, São Paulo, Brazil

Uldarico Malaspina
Department of Sciences
Pontificia Universidad Católica del Perú
Lima, Peru

ISSN 2211-1921 ISSN 2211-193X (electronic)
SpringerBriefs in Education
ISBN 978-3-319-97543-6 ISBN 978-3-319-97544-3 (eBook)
https://doi.org/10.1007/978-3-319-97544-3

Library of Congress Control Number: 2018952596

Foreword

The CANP project was set up to enhance and support representation and participation from many countries in the regions currently under-represented in ICMI, and as such is a critical activity of the ICMI. CANP was launched in 2009, and since then has been supported by three successive Executive Committees of ICMI. The evaluation of the CANP, has reported its overall success. It is with great pleasure that I welcome this publication of CANP work. It is one of a few, but growing publications that provides the international community with information and analysis of mathematics teacher education—a key component of mathematics education activity in a part of the world that typically is not well known.

The present volume is the third in the series of reports of the Capacity and Networking Project (CANP) of the International Commission on Mathematical Instruction (ICMI). It is the outcome of the intensive work that was carried out during and following CANP5 that was held in Lima, Peru for two weeks in February 2016. CANP5 brought together mathematics educators in the Andean Region (Ecuador, Peru, Bolivia) and Paraguay. The workshop engaged the many participants in a range of mathematical and mathematics education activities. A key component of the workshop was the four country reports on mathematics teacher education. This volume brings together this significant work and insights developed during the workshop across the delegates from participating countries. It is thus a very important resource for all working in mathematics teacher education internationally. The successes and challenges described in the book, while illuminating of conditions in under-resourced countries globally, are all part of the landscape that must inform our collective work as an organisation.

I take this opportunity to express the thanks of ICMI to all those who enabled CANP5—organisers, lecturers, participants—in the first instance, and following the workshop, all those who have made this volume possible. They are acknowledged individually and collective in the preface and introduction to this volume, and not repeated here. It nevertheless must be acknowledged that the volume is the outcome of a strong collaboration between all participants, the editors and other supports, not to mention those who worked on its translation into English for the benefit of the wider community. A volume such as this takes extensive work, all of which is done with generosity and commitment to the improvement of mathematics education worldwide. Specifically, our thanks to Yuriko Yamamoto Baldin and Uldarico Malaspina Jurado, for their leadership and their roles as editors of the volume and in the organisation of the workshop.

It is the strong hope of ICMI that through CANP volumes such as this, our understanding of mathematics education successes, needs and problems across different regions of the world will enable the advance our field in ways that are of the benefit of all.

Johannesburg, South Africa
July 2018

Jill Adler President of the International Commission on Mathematical Instruction

Preface

The Capacity and Networking Project (CANP), carried out by the International Commission on Mathematical Instruction (ICMI) of the International Mathematics Union (IMU) is one of most important outreach initiatives of ICMI-IMU to developing countries, supported by the Commission for Developing Countries (CDC) of IMU. The project is part of the response to the challenges posed in the UNESCO document "*Les défis de l'enseignement des mathématiques dans l'éducation de base*" (Current Challenges in Basic Mathematics Education), published in 2011, and, in the same year, the first CANP project was organized in Sub-Saharan region of Africa. The subsequent editions of CANP projects have been organized in regions determined to be in need of collaborative actions of educators and mathematicians, educational systems officials and teachers, with the goal of constructing sustainable regional networks to develop the local potential in facing the issues and challenges of the teaching and learning of mathematics.

The fifth CANP was carried out as CANP 5 for the Andean Region and Paraguay in Lima, Peru, during two weeks of February 2016, hosted by the Pontificia Universidad Católica del Perú (PUCP), with delegations from four countries: Ecuador, Peru, Bolivia from the developing Andean Region, and Paraguay. The organization of CANP 5 has sought the general objectives of CANP since its first edition, which is the improvement of the quality mathematics education in developing countries. The scientific program was developed by the members of the International Program Committee and the special invited experts from the international academic community. The main theme of CANP 5 had been previously chosen as "Initial and Continuing Mathematics Teacher Education", approved by the International Program Committee, and collectively agreed to by the country representatives to CANP 5 and the Local Organizing Committee. The discussions on the theme considered important subthemes such as *stimulus to mathematical thinking; curriculum of mathematics in the education at all levels; assessment; relations among mathematics, mathematics education, and the sciences; and a collaborative network among the participant countries*. These subthemes were addressed by the experts through lectures, video-conferences, hands-on workshops, and group discussions to support the final achievements of the CANP.

The country reports on the topic of Mathematics Teacher Preparation, elaborated following the guiding instructions of ICMI were important documentation to support the group discussions. The systematization of the format and the topics for the country reports was the contribution of CANP 2—Central America and the Caribbean, realized successfully in Costa Rica in 2012, allowing a comparative analysis of issues and challenges of mathematics education in the participating countries.

We see CANP 5 as the continuation of the initiatives of CANP 2. Both CANP served regions that share many of the same issues and challenges in mathematics education, as well as mutually familiar social and cultural contexts. These similarities have strengthened the conjunction of the networks REDUMATE (Red de Educación Matemática de America Central y El Caribe—Network of Mathematics Education of Central America and the Caribbean) of CANP 2 and CEMAS (Comunidad de Educación Matemática de America del Sur—Community of Mathematics Education of South America) of CANP 5, with the objectives of further developing joint activities and projects to improve the teaching and learning of mathematics in the regions.

It has been a great honor for us to be the coordinators and organizers of the CANP 5. We witnessed the growth of the community in the follow up meeting held in Guayaquil, Ecuador, in April 2017, hosted by Escuela Superior Politécnica del Litoral (ESPOL). In that meeting the final versions of the country reports were developed as academic texts whose translation into English constitute the chapters of this book.

Each chapter corresponds to the report on teacher preparation in each participant country under the responsibility of the country representative to CANP 5, Sonia Cordero for Bolivia, Margarita Martínez for Ecuador, Gabriela Gómez for Paraguay, and Augusta Osorio for Peru, to whom we express deep appreciation for their commitment to CANP 5 activities, and especially for the dedication to elaborate and to coordinate the country reports with the delegates of their country. We wish to extend our appreciation to all the country delegates.

The role of CANP 2 in supporting CANP 5 was remarkable in showing how to share the academic strengths and expertise of all who were involved, offering not only opportunities for participation in conferences and the advice/reviews from the mathematics education researchers, but especially the publication of articles in Spanish, the outcome from the country reports, in the scientific journal, *Cuadernos de Investigación y Formación en Educación Matemática*, published by Universidad de Costa Rica. We express our deepest appreciation to Angel Ruiz, the chief editor of *Cuadernos* and Coordinator of CANP 2, for invaluable support and help to CANP 5 from the start, sharing generously his knowledge and expertise, offering suggestions, guidance and, above all, his friendship and understanding.

Thanks are due to Patrick Scott for his superb work of translation of originals in Spanish into English and reviewing the language of whole text of this book, besides his participation in the CANP 5 activities with constant support that goes far beyond the execution of routine tasks.

We acknowledge the international lecturers Ferdinando Arzarello, Patrick Scott, Vicenç Font, Masami Isoda and Cesar Lau for their invaluable contribution to the scientific program of CANP 5 in Lima, generously participating out of ICMI-PUCP budget.

Our special thank you goes to the Pontificia Universidad Católica del Perú (PUCP) for its financial and logistical support in hosting the venue for the CANP 5, offering excellent infrastructure and support to the participants, during two weeks in February 2016. Our thanks are extended as well to the Escuela Superior Politécnica del Litoral (ESPOL) for the excellent conditions on its campus in Guayaquil, Ecuador, that permitted the work of the follow-up meeting of CANP 5, in April 2017. We also acknowledge CASIO Latin America for materials at the CANP 5 workshop held in Lima in 2016.

We would also like to thank Lena Koch, the ICMI/CDC Administrator, as well as the staff at Springer. Without their encouragement and support this publication would not have been possible.

São Carlos, Brazil
Lima, Peru
June 2018

Yuriko Yamamoto Baldin
Uldarico Malaspina

Contents

CANP 5—Andean Region and Paraguay (Lima, Peru, 1–12 February 2016)

Organization

International Commission on Mathematical Instruction (ICMI)
International Mathematical Union-Committee for Developing Countries (IMU-CDC)
Pontificia Universidad Católica del Perú (PUCP)

International Program Committee (IPC)

- Yuriko Yamamoto Baldin (Universidade Federal de São Carlos, Brasil, Chair—Liaison ICMI)
- Ángel Ruiz (Universidad de Costa Rica, Costa Rica)
- Christian Schaerer (Universidad Nacional de Asunción, Paraguay)
- José Miguel Contreras (Universidad de Granada, España)
- Luis Radford (Laurentian University, Canadá)
- Michèle Artigue (Université de Paris-Diderot, France)
- Roger Metzger (Instituto de Matemáticas y Ciencias Afines-IMCA, Perú)
- Salomé Martínez (Universidad de Chile, Chile)

Local Organizing Committee (LOC)

- Uldarico Malaspina Jurado (Pontificia Universidad Católica del Perú—IREM, Chair)
- Augusta Rosa Osorio Gonzales (PUCP, Representative of Peru)
- Estela Aurora Vallejo Vargas (PUCP, Perú)
- Gabriela Gómez Pasquali (Organización Multidisciplinaria de Apoyo a Profesores y Alumnos—OMAPA, Representative of Paraguay)
- Margarita Helena Martínez Jara (Escuela Superior Politécnica del Litoral—ESPOL, Representative of Ecuador)

- Olimpia Rosa Castro Mora (Ministerio de Educación, Perú)
- Rudy José Rosas Bazán (Sociedad Matemática Peruana, Perú)
- Sonia Cordero Cárdenas (Universidad Mayor de San Andrés—UMSA, Representative of Bolivia)

Invited International Lecturers—CANP 5

- Ferdinando Arzarello (Universitá di Torino, ex-ICMI President, Italy)—opening lecture
- Alan Schoenfeld (University of Berkeley, USA)—video conference
- Patrick Scott (New Mexico State University, CIAEM Vice President, USA)
- Jose Miguel Contreras (Universidad de Granada, Spain)—video conference
- Vicenç Font (Universidad de Barcelona, Spain)
- Angel Ruiz (Universidad de Costa Rica, ex-ICMI Vice President, CIAEM President, Costa Rica)
- Michèle Artigue (Université Paris-Diderot, ex-President ICMI, France)
- Luis Radford (Laurentian University, Vice president ICMI, Canada)
- Masami Isoda (University of Tsukuba, Japan)
- Cesar Lau (CASIO Latino America, Peru)

The editors wish to express special gratitude to Michèle Artigue for the definitive role in the development of the CANP project started during her Presidency in ICMI-IMU, and for all the contribution since the organization of first CANP in 2011. The organizers of CANP 5 heartfully acknowledge her continuous support, guidance and academic contributions throughout the journey to this date, generously sharing her outstanding academic experience and knowledge along with friendly understanding of the region.

Delegations of the Participating Countries—CANP 5

Bolivia

- Sonia Ximena Nivia Cordero Cardenas (Universidad Mayor de San Andrés—UMSA, Representantive of Bolivia in the LOC)
- Begoña Grigoriu (President of Sociedad Boliviana de Educación Matemática—SOBOEDMA)
- Gustavo Michel Garcia (Universidad Mayor de San Andrés—UMSA)
- María Antonieta Valenzuela (SOBOEDMA)
- Reynaldo Guzmán (Universidad Pedagógica "Mariscal Sucre")
- Victoria Mamani (Universidad Pedagógica "Mariscal Sucre")

Ecuador

- Margarita Helena Martínez Jara (Escuela Superior Politécnica del Litoral—ESPOL, Representantive of Ecuador in the LOC)
- Andrea Ayala (Universidad San Francisco de Quito—USFQ)
- César Trelles (Universidad de Cuenca)
- Eulalia Calle (Universidad de Cuenca—UC)
- Fernanda Espinoza (Ministerio de Educación)
- Fredy Rivadeneira (Universidad Técnica de Manabí—UTM)
- Janeth Valdiviezo (Escuela Superior Politécnica del Litoral—ESPOL)
- Monica Flores (Universidad de Especialidades Espíritu Santo)
- Nely Gonzalez (Universidad de Cuenca—UC)
- Paola Castillo (Universidad San Francisco de Quito—USFQ)
- Roxana Auccahuallpa Fernández (Universidad Nacional de Educación—UNAE)

Paraguay

- Gabriela Gómez Pasquali (Organización Multidisciplinaria de Apoyo a Profesores y Alumnos—OMAPA, Representantive of Paraguay in the LOC)
- Diana Giménez de von Lücken (Universidad Católica de Asunción)
- Jorge Daniel Mello Román (Decano de la Facultad de Ciencias Exactas y Tecnológicas—FACET, Universidad Nacional de Concepción)
- María Elisa Maidana de Giangreco (Instituto Superior de Educación)
- Osvaldo Vega (Politécnico y Facultad de Ciencias Exactas y Naturales, Universidad Nacional de Asunción)
- Richard Ramón Solís Argüello (Facultad de Ciencias Exactas y tecnológicas—FACET, Universidad Nacional de Concepción)
- Sergio Ayala (Organización Multidisciplinaria de Apoyo a Profesores y Alumnos—OMAPA)

Peru

- Augusta Osorio Gonzales (Instituto de Investigación sobre la Enseñanza de las Matemáticas IREM-PUCP, Representantive of Perú in the LOC)
- Carlos Manuel Sabino Escobar (Universidad Nacional de Tumbes)
- Carlos Torres Ninahuanca (M.Sc. Student-PUCP; High School Teacher)
- Cerapio Quintanilla Cóndor (Universidad Nacional de Huancavelica)
- Elizabeth Advíncula (IREM-PUCP)
- Estela Aurora Vallejo Vargas (IREM-PUCP, LOC)
- Giovanna Piscoya Ramírez (Dirección General de Educación Básica Regular—MINEDU)
- Jorge Enrique Quiroz Quiroz (Universidad Nacional de Educación Enrique Guzmán y Valle)
- Katia Vigo Ingar (Maestría en Enseñanza de las Masstemáticas, PUCP)
- Lilian Isidro Cámac (Dirección General de Educación Básica Regular—MINEDU)
- María Angelita Aredo Alvarado (Universidad Nacional de Piura)
- María del Carmen Bonilla Tumialán (Asociación Peruana de Investigación en Educación Matemática—APINEMA)
- Martha Rosa Villavicencio Ubillus (Dirección de Educación Intercultural y Bilingüe—MINEDU)
- Miguel Díaz Sebastián (Instituto Pedagógico Nacional de Monterrico)
- Norma Rubio Goycochea (Sección Matemáticas—PUCP)
- Olimpia Rosa Castro Mora (Sociedad Peruana de Educación Matemática, MINEDU, LOC)
- Rudy José Rosas Bazán (PUCP, Sociedad Matemática Peruana, LOC)

Acronyms

Chapter "Report on Mathematics Teacher Preparation in Bolivia"

DGFM	Dirección General de Formación de Maestros (Directorate General for Teacher Preparation)
ESFM	Escuelas Superiores de Formación de Maestras y Maestros (Teacher Preparation Colleges)
LLECE	Laboratorio Latinoamericano de Evaluación de la Calidad de la Educación (Latin American Laboratory for the Evaluation of the Quality of Education)
PEC	Práctica Educativa Comunitaria (Community Education Student Teaching)
PROFOCOM	Programa de Formación Complementaria (Program for Complementary Preparation)
SEP	Subsistema de Educación Superior de Formación Profesional (Higher Education Subsystem for Professional Preparation)
SOBOEDMA	Sociedad Boliviana de la Educación Matemática (Bolivian Mathematics Education Society)
TICs	Tecnologías de la Información y la Comunicación (Information and Communication Technology)
UAs	Unidades Académicas (Academic Units)
UMSA	Universidad Mayor de San Andrés (Higher University of San Andrés)
UNEFCO	Unidad Especializada de Formación Continua (Specialized Unit on Continuous Preparation)

Chapter "Report on Mathematics Teacher Preparation in Ecuador"

AACES	Consejo de Evaluación, Acreditación y Aseguramiento de la Calidad de la Educación Superior (Council for Evaluation, Accreditation and Assurance of the Quality of Higher Education)
BGU	Bachillerato General Unificado (General Unified Upper Secondary)
CES	Consejo de Educación Superior (Higher Education Council)
EGB	Educación General Básica (General Basic Education
ENES	Examen Nacional de Educación Superior (National Higher Education Examination)
ESPOL	Escuela Superior Politécnica del Litoral (Polytechnic Higher School of the Coast)
IESALC	Educación Superior Virtual en América Latina y el Caribe. (Higher Virtual Education in Latin America and the Caribbe)
INEC	Instituto Nacional de Estadísticas y Censos (National Institute of Statistics and Census)
INEVAL	Instituto Nacional de Evaluación Educativa (National Institute of Educational Assessment)
LOEI	Ley Orgánica de Educación Intercultural (Organic Law of Intercultural Education)
LOES	Ley Orgánica de Educación Superior (Organic Law of Higher Education)
MINEDUC	Ministerio de Educación (Ministry of Education)
SEDEM	Sociedad Ecuatoriana de Matemáticas (Ecuadorean Society of Mathematics)
SENESCYT	Secretaria Nacional de Educación Superior, Ciencia y Tecnología (National Secretariat of Higher Education, Science, Technology and Innovation)
SIPROFE	Sistema Integral de Desarrollo Profesional Educativo (Integrated System of Educational Professional Development)
SNNA	Sistema Nacional de Nivelación y Admisión (National Leveling and Admission System)
UC	Universidad de Cuenca (University of Cuenca)
UCE	Universidad Central del Ecuador (Central University of Ecuador)
UEB	Universidad Estatal de Bolívar (State University of Bolivar)
UNACH	Universidad Nacional de Chimborazo (National University of Chimborazo)
UNAE	Universidad Nacional de Educación (National Education University)

UNED	Universidad de Educación a Distancia de España (Distance Education University of Spain)
UNESCO	United Nations Educational, Scientific and Cultural Organization
USFQ	Universidad San Francisco de Quito (San Francisco University of Quito)
UTM	Universidad Técnica de Manabí (Technical Universtity of Manabí)
UTN	Universidad Técnica del Norte (Technical University of North)
UTPL	Universidad Técnica Particular de Loja (Private Technical University of Loja)

Chapter "Report on Mathematics Teacher Preparation in Paraguay"

BECAL	Programa Nacional de Becas de Posgrado en el Exterior para el Fortalecimiento de la Investigación, la Innovación y la Educación, "Don Carlos Antonio López" (National Program for Graduate Study Abroad to Strengthen Research, "Don Carlos Antonio López")
BID (IDB)	Banco Interamericano de Desarrollo (The Inter-American Development Bank)
CONACYT	Consejo Nacional de Ciencia y Tecnología (National Council for Science and Technology)
CPI	Cursillo Preparatorio de Ingreso (Entrance Preparatory Course)
CRE	Centro Regional de Educación (Regional Education Centers)
DGPE	Dirección General de Planificación Educativa (Directorate General of Educational Planning)
EEB	Educación Escolar Básica (Basic School Education)
EI	Educación Inicial (Preprimary)
EM	Educación Media (Upper secondary)
FACEN	Facultad de Ciencias Exactas y Naturales (Faculty of Exact and Natural Sciences)
FEEI	Fondo para la Excelencia de la Educación y la Investigación (Fund for Excellence in Education and Research)
FONACIDE	Fondo Nacional de Inversión Pública y Desarrollo (National Fund for Public Investment and Development)
GCI	Índice Global de Competitividad (Global Competitiveness Index)
IFD	Instituto de Formación Docente (Teacher Preparation Institutes)
ISE	Instituto Superior de Educación (Higher Institute of Education)
MEC	Ministerio de Educación y Ciencias* (Ministry of Education and Science)
	*Anteriormente llamado Ministerio de Educación y Cultura, se realizó el cambio de nombre en enero de 2017

OEI	Organización de Estados Iberoamericanos (Organization of Ibero-American States)
OMAPA	Organización Multidisciplinaria de Apoyo a Profesores y Alumnos (Multidisciplinary Organization for Support to Teachers and Students)
PIB (GDP)	Producto Interno Bruto (Gross Domestic Product)
PROCIENCIA	Programa para el Desarrollo de la Ciencia y Tecnología (Program for the Development of Science and Technology)
PRONII	Programa Nacional de Incentivo a la Investigación (National Program for Incentivizing Research)
SIEC	Sistema de Información de Estadística Continua (Continuous Statistic Information System)
SINAD	Sistema Nacional de Actualización Docente (National System of Teacher Development)
STP	Secretaría de Planificación Técnica (Secretariat of Technical Planning)
TERCE	Tercer Estudio Regional Comparativo y Explicativo (Third Regional Comparative and Explanatory Study)
TIC	Tecnologías de la Información y Comunicación (Information and Communication Technology)
TP	Teórico-Práctico (Theoretical-Practical)
UNA	Universidad Nacional de Asunción (National University of Asunción)
UNC	Universidad Nacional de Concepción (National University of Concepción)
UNESCO	Organización de las Naciones Unidas para la Educación, la Ciencia y la Cultura (United Nations Educational, Scientific and Cultural Organization)
WEF	Foro Económico Mundial (World Economic Forum)

Chapter "Report on Mathematics Teacher Preparation in Perú"

IMUNI	Instituto de Matemática de la Universidad Nacional de Ingeniería (Institute of Mathematics of the National University of Engineering)
INIDE	Instituto Nacional de Investigación y Desarrollo de la Educación (National Institute for Research and Development of Education)
IPEM	Instituto para la Promoción de la Enseñanza de las Matemáticas (Institute for the Promotion of Mathematics Teaching)
OEI	Organización de Estados Iberoamericanos (Organization of Ibero-American States)

PEN	Proyecto Educativo Nacional (National Education Project)
PUCP	Pontificia Universidad Católica del Perú (Pontifical Catholic University of Perú)
SECPANE	Peruvian-North American Cooperative Education

Introduction

The aim of this Introduction chapter is two-fold. The first is to present the book *"Mathematics Teacher Education in the Andean Region and Paraguay: A Comparative Analysis of Issues and Challenges"* as a contribution to the series of Springer Briefs in Education that consist of the publication of ICMI-IMU CANP project reports on Teacher Education in developing countries. In summarizing the information presented by the participants of CANP 5, the authors of this introductory chapter consider how the struggles to construct a network community of participating countries have evolved from the first discussions of educational issues to a production of an organized academic text on the subject, which permits a comparative analysis of common issues as well as cultural differences. Such an analysis is a valuable source for further research on Mathematics Education in the developing countries of South America, as well as for decision making by educational policy stakeholders. The chapters of this book are revised and improved texts based on draft reports presented at CANP 5, and authored by the delegations of each participating country: Bolivia, Ecuador, Paraguay and Peru. They have been translated from the originals in Spanish.

The second objective is to summarize the critical reading of these chapters, as editors of the book and coordinators of the CANP 5 workshop realized in 2016, in Lima, Peru. Our comments in this regard, as insiders of the context of Mathematics Education in South America, reflect an analysis of common issues and differences, as well as the accomplishments since the CANP 5 workshop and the challenges for the future.

We comment briefly on the IMU/ICMI CANP project to situate the rationale for CANP 5, the 5th edition of this project, and its strategic importance to the improvement of Mathematics Education in four countries of South America facing significant educational challenges. Below we present each chapter, highlighting its main characteristics that allow a comparison with those of other countries. We offer a critical analysis of our findings that will be used as basis for the final considerations on accomplishments and challenges in the Conclusions chapter of the book.

CANP 5 and Mathematics Teacher Preparation

The "Capacity and Network Project" (CANP) is a major project of the International Mathematics Union (IMU) carried out by its International Commission on Mathematical Instruction (ICMI), since 2011, directed to developing countries to enhance their existing capabilities for meeting the challenges in Mathematics Education. (https://www.mathunion.org/icmi/activities/developing-countries-support/capacity-networking-project-canp).

By organizing workshops in selected regions of developing world, "the project provides mathematics teacher educators … with enhanced mathematical and pedagogical expertise through capacity and networking building." (Koch, 2017). This reference is a very important document on the evaluation of the CANP project that describes clearly the evolution of the project from its first edition CANP 1 to CANP 5 in the Andean Region and Paraguay that is the subject of this book. Throughout the process of various CANP, we have learned the different cultural contexts and educational policies, as well as the different levels of the most relevant issues in developing Mathematics Education. In this process the issue of ***Mathematics Teacher Preparation*** has been determined to be the most important topic in the activities and discussions, because the quality and capacities of school mathematics teachers impact the whole dimension of the mathematics education to foster the development of a country.

The region for CANP 5 has been chosen strategically in South America with the hope to join with the network REDUMATE that resulted from CANP 2, Central America and the Caribbean, to strengthen the networking capabilities as well as to take advantage of Spanish as a common language (Fig. 1).

Three countries of the Andean Region (Bolivia, Ecuador and Peru), as well as Paraguay participated in CANP 5. It was organized with the goal of improving the quality of mathematics education in these countries, and was realized in 2016, in Lima, Peru. The ultimate objective of the CANP 5 was the construction of a network of collaborators from participating countries to facilitate the sharing of experiences, and to open the doors to the internationalization of each country's efforts by participating in events to become part of the world community. The scientific program of CANP 5 focused strongly on Mathematics Teacher Preparation issues and challenges among other high-quality inputs from participating international academics to bring valuable resources into the Mathematics Education of the region, as already pointed out by ICMI as the core subject of the CANP project.

Therefore, the theme of "**Initial and Continuing Teacher Education**" led and permeated the discussions during CANP 5 activities that would progressively be worked out to converge into decisions/actions for creating a collaborative network. To support the discussions, the delegations of each country had brought a "country report" with answers to questions about "Mathematics Teacher Education: initial and continuing", following a guiding script from CANP-ICMI. The first reports

Fig. 1 Region of CANP 5 highlighted on South America map. *Source* (public domain): (https://www.cia.gov/library/publications/the-world-factbook/graphics/ref_maps/political/pdf/south_america.pdf)

were then technical reports to motivate the discussions and encourage comparisons between the special issues and common difficulties, understanding the regional capabilities and political issues in the teaching profession, and to foresee the challenges in constructing a sustainable network that could be linked to existing academic organizations and networks, notably the REDUMATE of CANP 2.

The collaborative work in the nascent community of CANP 5 has led to convert the technical reports discussed in the 2016 workshops into academic texts that now comprise a cohesive text that we believe will be very useful for various analyzes and research on mathematics education in the region. Some of the authors did not have previous significant experience in producing academic style texts, and the process has taken time. The work was completed after a follow-up meeting with leading persons from the participating countries, that was held in Guayaquil, Ecuador, in April 2017, with support from ICMI. This is a very meaningful achievement for CANP 5.

Commenting on the Chapters

We present brief comments on each chapter in the critical perspective of the editors, to distinguish the characteristics of each country. Following the orientation of ICMI, as considered in the first draft country reports, the basic structure of each chapter consists of the Historic Context of Teacher Preparation; the Structure of the Educational System (which may include the Standards and Laws); the Structure for Initial Preparation of Mathematics Teachers and the Contents of its Curriculum; the Continuous Preparation of Teachers; Careers in Teaching; Mathematics Education Research; Strengths, Weaknesses, Threats and Challenges.

The chapters are in alphabetical order by country names.

First Chapter: Report on Mathematics Teacher Preparation in Bolivia
This chapter does not present the historic background of the education system in that country before 1952, although we are well aware of Bolivia's rich cultural heritage that gives a context and important elements for explaining the actual state of Mathematics Education in the country. The chapter shows important aspects of the educational reality of Bolivia and helps to better understand its problems in a Latin American context. We consider the inclusion and participation of Bolivia in the CANP5 network very important, as a point of support to develop their own capacities to improve the quality of their mathematical education and to face the challenges of the future. Both the Teacher Education and the conception of the Curriculum of Basic Education present particularities that distinguish Bolivia from other countries of the region, with keen attention to an inclusive ethnic education, highlighting bilingual education and native culture. The attention to bilingual education and ethnic issues in Mathematics Education is present in all the other chapters. It is interesting to know the phenomenon of apparent oversupply of school teachers, due to the recruiting system policy. Yet the authors point to the lower achievement indices in mathematics education caused by shortcomings in teacher preparation, implying that continuing education is vital in seeking changes in education.

Second Chapter: Report on Mathematics Teacher Preparation in Ecuador
This chapter describes thoroughly the main topics of the basic structure with comments and critiques. It is the translation of an adapted text that evolved from the original in Spanish, published in the academic journal, *Cuadernos de Investigación*

y Formación en Educación Matemática, of the University of Costa Rica. It is interesting to learn the history of educational institutions in charge of Teacher Education, which were part of two subsystems, one for urban areas and another for rural areas, reflecting the history of colonization of the country. The evolution of the structures of the educational system is described by a sequence of laws, and the curricular guidelines as a consequence of the evolution related to the profiles of students exiting from the educational system, that includes the distinction of Integrative Curricular Processes from Learning Processes for Basic Education and Upper Secondary. This feature distinguishes the Curriculum of Ecuador and is not found in the other countries. The authors offer a critical analysis based on the topics of their report, and emphasize the current challenges with respect to the preparation of mathematics teachers. They point to the problem of weaknesses in the higher education accreditation system that allows some institutions to graduate future elementary or high school mathematics teachers without the knowledge and skills necessary to be successful in schools. This fact could be compared with the situation of other countries to propose possible future initiatives.

Third Chapter: Report on Mathematics Teacher Preparation in Paraguay
This Chapter reports very critically the situation of Mathematics Education in Paraguay, describing the challenges faced to improve the level of learning mathematics by students, linking the current situation to the weak preparation and professional development of mathematics teachers, thus discussing issues and challenges also found in the other chapters. The discussion considers the main topics of the basic structure of the education system, and is often very critical of the lack of development of academic knowledge and pedagogical skills of teacher educators who prepare in-service teachers. There is also a proliferation of low quality Teacher Preparation Institutes, that can be compared to the situation of teacher oversupply in Bolivia. The report points to a strong need for a research environment in the country, that could influence decisions on educational matters. This chapter contributes to the comparative analysis sought by this book, especially in the discussion on the challenges faced in the present and in the future.

Fourth Chapter: Report on Mathematics Teacher Preparation in Perú
This Chapter brings a thorough report about the issues of Mathematics Teacher Education, in an academic style that facilitates the reading and analysis. The text is also the translation of an adapted version published in *Cuadernos de Investigación y Formación en Educación Matemática*. Starting from a clear historical context of Teacher Preparation in Peru up to the current situation of the Educational System, this chapter discusses systematically each of the topics of the basic structure, always grounded in official documentation accompanied with critical comments. The tables with quantitative data about the offering of Mathematics Courses and Mathematics Teaching Courses for each level of education system are good references to support quality analysis about the formation of mathematics teachers that follow in the text. Along with the description of data the authors do not omit their critical interpretation about the details, showing the quality level of the research environment that exists in Mathematics Education in Peru. Though this observation points to the

strength of the Peruvian academic community, it is noticeable that the weaknesses and challenges indicated at the end of chapter have many commonalities with the other CANP 5 countries that contribute to a rich discussion of issues and challenges in the region.

The chapter on Conclusions in the book summarizes the analysis of the editors concerning the findings and reflections on the challenges discussed at CANP 5.

Yuriko Yamamoto Baldin
Uldarico Malaspina

Reference

Koch, L. (2017). *Evaluation of the capacity and network project (CANP) of the international commission on mathematical instruction (ICMI) full report.* Recovered from https://www.mathunion.org/icmi/activities/developing-countries-support/capacity-networking-project-canp

Report on Mathematics Teacher Preparation in Bolivia

Sonia Ximena Cordero, Begoña Grigoriu, María Antonieta Valenzuela, Reinaldo Guzmán, Victoria Mamani, Gustavo Michel and Humberto Giácoman

Y. Y. Baldin and U. Malaspina (eds.), *Mathematics Teacher Education in the Andean Region and Paraguay*, SpringerBriefs in Education,
https://doi.org/10.1007/978-3-319-97544-3_1

Introduction

The Plurinational State of Bolivia is located in South America. It has a population of 11.4 million and a surface area of 1,098,581 km^2. It is divided into nine departments. Sucre is the capital, while La Paz is the seat of government. (Muñoz Reyes, 1980)

S. X. Cordero (✉) · G. Michel
Universidad Mayor de San Andrés, La Paz, Bolivia
e-mail: sonia_cordero_de_r@yahoo.com

G. Michel
e-mail: don_gus13@yahoo.es

B. Grigoriu · M. A. Valenzuela
Sociedad Boliviana de la Educación Matemática (SOBOEDMA), Cochabamba, Bolivia
e-mail: bgrigoriu@gmail.com

M. A. Valenzuela
e-mail: antonietav1@hotmail.com

R. Guzmán
Universidad Pedagógica, Potosí, Bolivia
e-mail: naldo_m3@hotmail.com

V. Mamani
Universidad Pedagógica, Chuquisaca, Bolivia
e-mail: vmamani67@gmail.com

H. Giácoman
Sociedad Boliviana de la Educación Matemática (SOBOEDMA), La Paz, Bolivia
e-mail: betogiacoman@hotmail.com

Y. Y. Baldin and U. Malaspina (eds.), *Mathematics Teacher Education in the Andean Region and Paraguay*, https://doi.org/10.1007/978-3-319-97544-3_1

History of Teacher Preparation

Teacher preparation in Bolivia has had three major changes: the first with the land reform of 1952, the second with the 1994 educational reform law and finally with the enactment of the Avelino Siñani—Elizardo Perez law in 2006.

Initially teacher preparation was elitist, only preparing teachers for urban areas. In the 1930s the State promoted rural education. Siñani together with Elizardo Pérez founded the Ayllu of Warisata School. It was a model for integrating the school with the community (Ministerio de Educación, 1999).

After the National Revolution in 1952, teacher preparation was divided into rural and urban.

- To prepare teachers for urban elementary and secondary schools there were Normal Schools in Sucre and La Paz which offered both content and pedagogical courses, as well as student teaching.
- At eight rural Normal Schools, future teachers were presented basic academic content, but the greatest time was dedicated to *professional preparation*, agriculture, rural industry, hygiene and sanitation.

In 1994 the Education Reform Law 165 was passed. The Normal Schools became Higher Normal Institutes and were administered by the universities. At the national level, there were six Institutes for the urban areas in charge of preparing teachers for pre-school, elementary and secondary. In the rural areas, there were 12 Institutes that prepared elementary teachers in bilingual intercultural education (Republica de Boliria 1994).

In July of 2006, the National Education Congress proposed the "Avelino Siñani—Elizardo Pérez" Law. It took the Higher Normal Institutes out from under the control of the universities and renamed them as "Teacher Education Colleges" (*Escuelas Superiores de Formación de Maestras y Maestros*-ESFM) (República de Bolivia 2006).

Current Structure of the Educational System

The Plurinational Educational System is divided into three subsystems (Table 1).

Regular Education

The levels of education are defined by the grades in the schools:

- Early Childhood Level: Pre-Kinder and Kinder, Pre-Primary in Family Community (5–6 years old).
- Primary Elementary Level: Grades 1–3. Elementary Communal Vocational Education (6–8 years old).

Table 1 Structure of the Plurinational Educational System

Subsystem	Scope	Areas	Levels	Ages (years)
Regular Education	Pre-Primary in Family Community		In the home	0–4
			Pre-Primary	5
	Elementary Communal Vocational	Basic Vocational	Grades 1–5	6–10
		Advanced Vocational	Grades 6–8	11–15
	Secondary Communal Productive		Grades 9–12	14–18
Alternative and Special Education	Alternative Education	Adults		
		Permanent		
		Literacy and Post-literacy		
	Special Education	Disabilities		
		Learning difficulties		
		Extraordinary talent		
Higher Education and Professional Preparation	Teacher Preparation			
	Technical and Technological Preparation			
	Artistic Preparation			
	University Preparation			

Source (Ministerio de Educación, Cultura y Deportes de Bolivia, 2015)

- Upper Elementary Level: Grades 4–6. Elementary Communal Vocational Education (9–11 years old).
- Lower Secondary Level: Grades 7–8. Communal Advanced Vocational Education (12–14 years old).
- Upper Secondary Level: Grades 9–12. Communal Productive Education (14–18 years old).

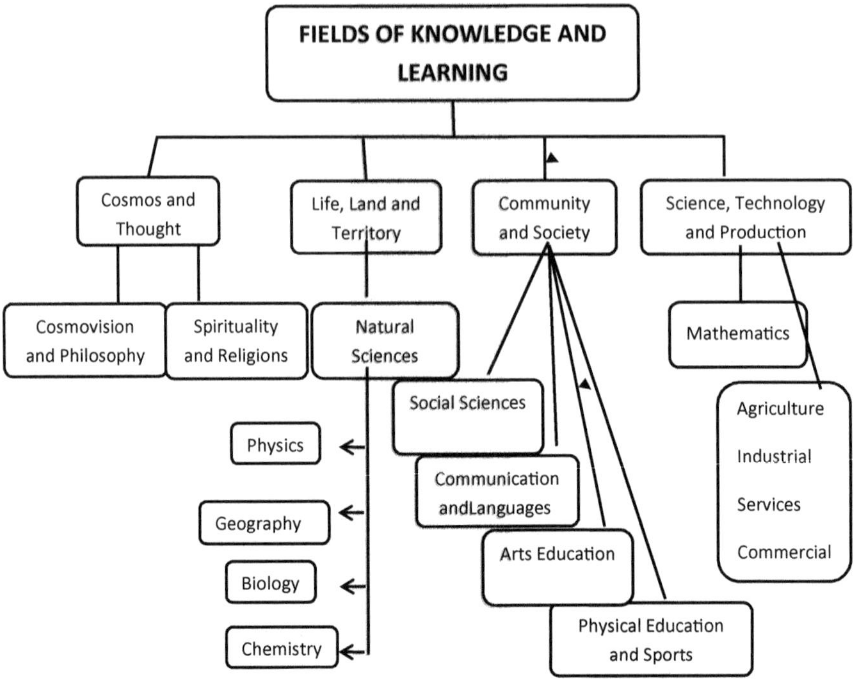

Fig. 1 Fields of Knowledge and Learning. *Source* Ministerio de Educación, Cultura y Deportes de Bolivia (2010a, b)

School subjects are classified as Fields of Knowledge and Learning that are divided into four areas:

- Cosmos and Thought
- Life, Land and Territory
- Community and Society
- Science, Technology and Production.

Mathematics is found in area of Science, Technology and Production, with the following components: Geometry, Arithmetic, Algebra, Calculus and Statistics.

All subjects are developed along four dimensions: Being, Knowing, Doing and Deciding.

- *Being*, develops principles of community life.
- *Knowing*, develops knowledge, sciences and technologies.
- *Doing*, develops the practice and the production as social manifestation.
- *Deciding*, develops a person's political area so that they are able to respond to life's challenges (Fig. 1).

Mathematics is required at all grade levels, with more dedicated hours at the secondary level: 5 academic hours per week for Grades 7–10, and 4 h in Grades

11 and 12. Therefore, there are 120 h of Mathematics in the 6 years of secondary education.

Upon completing the 13 years of study, students graduate as:

- Graduate (*Bachiller*) in Humanities, or
- Technical-Humanistic Graduate.

Preparing Teachers

General Organization: Structures and Participants

The preparation of teachers in Bolivia is within the Higher Education Subsystem for Professional Preparation (*Subsistema de Educación Superior de Formación Profesional-SEP*) which is charged with preparing teachers for Regular Education, as well as Alternative and Special Education.

Teacher preparation is developed in three stages: Initial, Continuing and Graduate.

Initial Preparation

The preparation of future teachers is supervised by the Ministry of Education through the Directorate General for Teacher Preparation (*Dirección General de Formación de Maestros-DGFM*). The *DGFM* is in charge of the Teacher Education Colleges (*ESFM*) and the Academic Units (*Unidades Académicas-UAs*) that exclusively offer academic initial teacher preparation (Art. 36), República de Bolivia, 2006).

The initial preparation of future teachers is a five-year program. It concludes with the awarding of a bachelor's degree (*licenciatura*) as called for in article 35 of Education Law 070 "Avelino Siñani—Elizardo Pérez" (República de Bolivia, 2006).

The main route into the *ESFM* and *UA*s is defined by an admission's examination. When the Ministry of Education determines that there are needs for teachers in certain areas an announcement is published. Currently, there is an oversupply of Mathematics teachers. The only areas being solicited are Pre-Primary, Music, Fine Arts and English.

Entrance to initial teacher preparation programs is possible in via four modalities:

- Admission test
- Achievement and being a member of a rural native indigenous group
- Secondary school graduates who are outstanding athletes
- Secondary school graduates with disabilities.

There are public announcements for only the first two modalities: 80% of entering students are based on the entrance examination and the remaining 20% are given to indigenous graduates of secondary schools who had the highest grades during their

Table 2 National Quotas for the *ESFM*

Year	Quotas	Entrance by Examination 80%	Entrance of Indigenous Peoples 20%	Quotas in Mathematics	Areas Convened
2011	3500	2800	700	0	Physical Education Music Fine Art English Philosophy
2013	4305	3444	861	70	Pre-Primary Physical Education Music Fine Art English Philosophy Social Sciences Natural Sciences **Mathematics**
2015	3520	2816	704	0	Pre-Primary Physical Education Music Fine Art English
2016	2875	2300	575	0	Physical Education Music Fine Art English Philosophy

Source The authors

four years of secondary school. Entrants from the other two modalities are for special cases and there is not a set quota (Table 2).

In December of 2015, the first promotion of 5688 students received the title of Bachelors (*Licenciatura*) under the new education law. They are all required to work for two years in the provinces. The two top graduates of each *ESFM* and *UA* are given an incentive of a guaranteed job in school. The rest are gradually accommodated, according to needs.

Elementary Teacher Preparation

"Classroom Teachers" are prepared to teach various subjects. Only Religion, Physical Education and Music are taught by teachers prepared specifically for those disciplines.

Elementary teacher preparation is structured as indicated in Table 3.

The Mathematics courses for Elementary teacher preparation are shown in Table 4.

Table 3 Structure of Elementary teacher preparation

Areas of preparation	Instruction hours	Percentage (%)
General preparation (Includes Pedagogy, 80 h in year 2; and Mathematics Teaching, 160 h in year 4	1760	33
Specialized preparation in other fields	3080	58
Preparation in Sciences, Technology and Production (Mathematics)	440	9
Total Instruction Hours	**5280**	**100**
Community Education Student Teaching	600	

Source The authors

Table 4 Mathematics course in Elementary teacher preparation

Year in the program	Course	Instruction hours
Year 2	Mathematics in Community Cultures	80
Year 3	Geometry, Its Shapes and Relations	120
Year 4	Algebra as a Process from the Concrete to the Abstract	120
Year 5	Mathematics Teaching	120

Source The authors

Table 5 Structure of the preparation of Secondary Mathematics teachers

Areas of preparation	Instruction hours	Percentage (%)
General preparation (Includes Pedagogy, 80 h in year 2; and Mathematics Teaching, 160 h in year 4)	1760	33
Specialized preparation	2160	41
Educational Research	1360	26
Total Instruction Hours	**5280**	**100**
Community Education Student Teaching	600	

Source The authors

Secondary Teacher Preparation

After five years of study students are awarded the title of Bachelors (*Licenciatura*) in Secondary Productive Community Education with a major in Mathematics.

The structure of the preparation of Secondary Mathematics teachers in indicated in Table 5.

Table 6 presents the curriculum for the Bachelors in Secondary Productive Community Education with a major in Mathematics.

Table 6 Curriculum for the Bachelors in Secondary Community Productive Education with a major in Mathematics

Principle	Key elements	Fields of Knowledge and Learning	General Preparation				Specialized Preparation					
			Year 1		Year 2		Year 3		Year 4		Year 5	
			Courses	Hrs	Courses	Hrs	Courses	Hrs	Courses	Hrs	Courses	Hrs
Decolonization, Intracultural, Intercultural and Plurilingual Productive and Territorial Technical Scientific, Technological and Artistic	1. Intracultural, Intercultural and Plurilingual Education 2. Education in Socio-Communal Values 3. Education for Production 4. Education for Coexistence with Nature, and Community Health	Cosmos and Thought	Worldviews and Philosophies	120					Learning Difficulties and Curricular Adaptations I	80	Learning Difficulties and Curricular Adaptations Ii	80
			Preparation in Socio-Communal Values	80					Teaching Mathematics	160		
			Psychology, Human Development and Educational Change	80	Special Education I	80	Special Education II	80				

(continued)

Table 6 (continued)

Principle	Key elements	Fields of Knowledge and Learning	General Preparation				Specialized Preparation					
			Year 1		Year 2		Year 3		Year 4		Year 5	
			Courses	Hrs	Courses	Hrs	Courses	Hrs	Courses	Hrs	Courses	Hrs
		Society and Community	Psycho-pedagogical Theories	80	Pedagogy and Curriculum	80						
			Educational Manage-ment and Planning	80								
			The State and Education	160	Socio-Politics and Decolonization	160						
			Spanish Language Workshop I	80	Spanish Language Workshop II	80						
			Indigenous Language Workshop I	80	Indigenous Language Workshop II	80	Indigenous Language Workshop III	80	Indigenous Language Workshop IV	80	Indigenous Language Workshop V	80
		Life, Land and Territory			Family Community Health	80						
		Science, Technology and Production			Fundamentals of Arithmetic and Algebra Applied to Production	120	Algebra, Language and Concrete to Abstract Thought	160	Linear Algebra: Design and Production	160	Operations Research	120
											Topology	80

(continued)

Table 6 (continued)

Principle	Key elements	Fields of Knowledge and Learning	General Preparation				Specialized Preparation					
			Year 1		Year 2		Year 3		Year 4		Year 5	
			Courses	Hrs	Courses	Hrs	Courses	Hrs	Courses	Hrs	Courses	Hrs
					Geometry of our Cultures and Diversity	120	Geometry and Trigonometry of Space (Relations in Harmony with the Cosmos)	160			Financial Mathematics and Profitable Product Enterprises	120
							Calculus in R2 Applied to Technology	160	Calculus in R, Harmonic Relation and equilibrated with Production	200	Differential Equations and Complex Variables	200
							Descriptive and Inferential Statistics	160			Complementarity of Math with other Sciences	120
							Epistemology of Mathematics	80	Numerical Methods	160	Simulation and Mathematical Models	160
			ICT and Education I	80	ICT and Education II	80			Applied Informatics	120		
			Educational Research and Production of Knowledge I	120	Educational Research and Production of Knowledge II	160	Educational Research and Production of Knowledge III	160	Educational Research and Production of Knowledge IV	160	Educational Research and Production of Knowledge V	160
				960		1040		1040		1120		1120

Source Ministerio de Educación (2015)

Table 7 Distribution of time for the Community Education Student Teaching (*PEC*)

Stages	Phases			Hours
	Management	Community	Educational Institutions	
Year 1	Permanent-Indigenous towns, urban and/or rural communities			40–2 weeks
Year 2	Permanent-Indigenous towns, urban and/or rural communities			60–3 weeks
Year 3	Permanent-Indigenous towns, urban and/or rural communities			100–5 weeks
Year 4	Permanent-Indigenous towns, urban and/or rural communities			160–2 months
Year 5	Permanent-Indigenous towns, urban and/or rural communities			240–3 months
Total Hours				**600**

Source Ministerio de Educación (2012)

Community Education Student Teaching (*Práctica Educativa Comunitaria-PEC*)

Community Education Student Teaching (*PEC*) is developed beginning in Year 1. It is realized through student teaching in various educational institutions where the future teachers internalize and put pedagogy and their subject into practice.

The *PEC*, integrated with Educational Research and Production of Knowledge, is for a total of 600 h, distributed gradually (see Table 7).

Continuing Preparation of Teachers

The Specialized Unit on Continuing Preparation (*Unidad Especializada de Formación Continua-UNEFCO*) is in charge of the teacher professional development. The *UNEFCO* is considered to be a decentralized unit of the Ministry of Education and offers courses based on teacher demand.

There is also another unit in what the Ministry of Education call "Teacher Network" that provides online professional development for registered in-service teachers. The Teacher Network is currently offering courses on the use of mathematical software such as GeoGebra, MathLab and Latex.

Until 2009, the Bolivian Mathematics Education Society (*Sociedad Boliviana de Educación Matemática-SOBOEDMA*) regularly offered courses for teachers. However, with the passage of the Avelino Siñani—Elizardo Pérez Law, that work had to be suspended because only governmental entities can now offer professional development that has a curricular content.

One of the most popular programs for continuing preparation is the Program for Complementary Preparation (*PROFCOM*). It is financed by the Ministry of Education to raise the academic background of all teachers to the level of a bachelors

Table 8 Mathematics courses in the fourth (last) semester of *PROFOCOM*

For elementary teachers	For secondary teachers
Natural numbers	Rational numbers
Measurement	Algebra
Whole numbers and fractions	Trigonometry and analytic geometry
"Spacialities (*Espacialidades*)" and geometry	Economic and financial Mathematics

Source (Ministerio de Educación, 2016)

(*licenciatura*) degree. This has been seen as necessary since many teachers either had a teaching degree at a lower level or a degree from some other discipline (Ministerio de Educación, 2016).

The analytic content of the *PROFOCOM* courses is quite ideological with less time given to the actual subject matter. There are four courses a semester and a total of four semester.

Table 8 shows the courses that are presented during the fourth semester of the program for Mathematics.

PROFOCOM has developed in three phases and the first phase began in the first semester of 2012. It was for elementary teachers in Grades 1, 2 and 3, and secondary teachers in the first 2 years. The second phase began in the second semester of 2013 with participation from elementary teachers from Grades 3 and 4, and secondary teachers from the rest of the grades. The third phase, beginning in the first semester of 2015, was opened to any teachers who did not have a Bachelors degree in education.

Because *PROFOCOM* was sponsored by the Ministry of Education and the degree confirmed implied a higher salary, a significant number of teachers registered: 135,000 as of 2014. The success of the first three phases led to a fourth phase in the second semester of 2016 that was opened to all teachers who still did not have a Bachelors degree.

The Mathematics Education needs of Bolivia are many. Bolivia is a country with the one of the lower achievement indices in Mathematics Education. The situation is due in great part to shortcomings in teacher preparation. Therefore, continuing teacher preparation is vital to attain changes in education.

Graduate Preparation

The institution responsible for the graduate education of teachers is the "Mariscal Sucre" Pedagogical University which began operations in 2015. It depends directly upon the Ministry of Education and has outreach throughout the country. It has offered four Masters degrees each lasting five semesters culminating with a final project:

Table 9 Distribution of teachers by departments in Bolivia

Department	Percent of in-service teachers (%)	Department	Percent of in-service teachers (%)
La Paz	28	Oruro	6
Santa Cruz	21	Beni	5
Cochabamba	17	Tarija	5
Potosí	10	Pando	1
Chuqisaca	7		

Source Ministerio de Educación (2017)

Table 10 Distribution of teachers by area in Bolivia

Area	Percent of in-service teachers (%)
Rural	40
Urban	37
Provincial	23

Source Ministerio de Educación (2017)

- Masters in Community Productive Secondary Mathematics Education
- Masters in Community Productive Secondary Physics/Chemistry Education
- Masters in Teacher Preparation Policy
- Masters in Intra/Intercultural Plurilingual Education.

The Masters in Mathematics Education has 133 registered students which is a small percentage of the total of secondary mathematics teachers. The students come from most of the departments of the country. The program began in October of 2015 and will finish in December of 2017.

The Pedagogical University also offered a shorter Specialization in Secondary Mathematics Education. It ran from April to November of 2016 with 292 registered.

Current Teacher Situation

In Bolivia there are over 135,000 teachers. The greatest number (28%) work in the department of La Paz (Table 9).

Bolivian teachers work in urban, rural and provincial areas. Provincial areas are those rural areas that have basic services such as water and electricity (Table 10).

The salary schedule for teachers has seven levels. Beginning teachers start at the fifth level and work their way up to the zero level and merit level (Table 11).

Table 11 Distribution of teachers by salary category in Bolivia

Salary	Percent of in-service teachers (%)	Salary	Percent of in-service teachers (%)
Fifth	37	First	6
Fourth	15	Zero	6
Third	10	Merit	13
Second	9	Temporary	4

Source Ministerio de Educación (2017)

Notable Recent Action

In 2016, the Bolivian government made the decision to participate in a program of international assessment called the Latin American Laboratory for the Evaluation of the Quality of Education (*Laboratorio Latinoamericano de Evaluación de la Calidad de la Educación-LLECE*). The first of those assessment were planned to be implemented in 2017 at the elementary level.

In 2010 professors in the Faculty of Engineering and the Faculty of Pure and Natural Sciences at the Higher University of San Andrés (*Universidad Mayor de San Andrés-UMSA*) moved to unify the existing scientific olympiads. The result was the Plurinational Bolivian Student Scientific Olympiad. The disciplines include Mathematics, Physics, Chemistry, Astronomy and Astrophysics, Biology, Informatics, Robotics and Geography.

The Olympiads for Mathematics, Physics, Chemistry, Astronomy and Astrophysics participate at the international level. The first to begin international competition was Mathematics in 1985. With the growth of these olympiads and the support of the ViMinistry of Education, student and parent expectations have grown. Therefore, teachers have become conscious of the material that is presented in the Olympiads and have searched for new teaching methods in the science disciplines, hoping that their students will be classified as the best in the country.

The Bolivian Mathematics Education Society (*Sociedad Boliviana de Educación Matemática-SOBOEDMA*) was created in 1995 to optimize the teaching of Mathematics. Since its creation, *SOBOEDMA* had searched for ways to integrate Mathematics into daily life and to promote the exchange of education research.

Initiatives do exist, independent of government policy, that have served to motivate teachers and students to improve the teaching of Mathematics by applying new methodologies.

Teacher Preparation and Research

The higher normal schools had offices of research and teacher preparation, when they were administered by the universities. Those offices were dedicated to planning, executing and motivating scholarly work with the participation of the faculty in research and professional development activities.

The research that was carried out in different areas was often published in books or photocopies. The most common topics were related to pedagogy and curriculum.

With the passage of the current education law "Avelino Siñani—Elizardo Pérez" and the establishment of PROFOCOM, some of the PROFOCOM end of course projects were in Mathematics. However, since there is not currently preparation of new Mathematics teachers, work in that area is practically suspended.

Strengths, Weaknesses, Opportunities and Threats

Strengths

- Teacher improvement is possible through graduate study and professional development courses.
- The educational system is offering better working conditions in terms of infrastructure and technology with respect to other professions.
- The majority of teachers, in all disciplines and at all levels, receive a computer to use in their teaching.
- There are many people who would like to study to become Mathematics teachers.
- The Plurinational Bolivian Student Scientific Olympiad presents a priority on the scientific and technological preparation of students. The disciplines include Mathematics, Physics, Chemistry, Astrophysical Astronomy, Biology, Informatics, Robotics and Geography. Students and teachers are thus improving their academic levels.

Weaknesses

- The performance of teachers is not evaluated and not considered in salary decisions.
- There is no follow-up system for evaluation of teacher preparation to guarantee updating of teacher knowledge.
- There are no up to date statistics on where teachers received their preparation nor on the work they are doing.
- Teacher placement on the salary scale is not well-defined.
- Union membership is obligatory.
- There are no mechanisms for competition among certified teachers thus the impetus to improve is stifled.
- The examination used in teacher promotion only measures administrative, pedagogical and philosophical aspects, without considering knowledge of the discipline in which they are teaching.

- Given the politicization of teachers, many academic and technical decisions are tainted with politics.
- Classroom teaching is based on routine memorization.
- Technology is not used in teaching Mathematics.
- The Mathematics Olympiad is perceived to be additional workload for teachers that is not well-remunerated.
- Teaching practice is still very repetitive and procedural, without developing mathematical thinking.
- Preparation for teaching Geometry is very weak.
- Teaching materials in use still prioritize Arithmetic and Algebra separate from Geometry.
- Formal symbolic logic has been removed from the curriculum, but is maintained philosophically and that complicates student understanding.
- There is not curricular coherence from grade to grade.
- Few teachers attend course on the use of technology in teaching.
- Mathematics teacher preparation has been eliminated by the *ESFM*.
- There are no research projects.
- There is an excessive ideological load in the curriculum for future teachers.
- Teachers in rural area often miss classes.

Opportunities

- There are entities, like the Teacher Network, where instructional materials and online courses are available for teachers who can use technological resources.
- The Pedagogical University offers a Masters in Mathematics for secondary in-service teachers.
- Teachers have the opportunity to accompany Bolivian student delegations to scientific competitions with the official title of Teacher Tutor.

Threats

- There is an oversupply of teachers.
- There are few opportunities to work in urban areas.
- There is a need for more specialized Mathematics teachers in rural areas.
- There is a lack of responsibility and commitment among teachers.
- Low salaries.
- Mathematics is mystified as terrifying.
- Some teachers maintain a threatening attitude toward their students.
- Recent education graduates are demoralized by the lack of job opportunities.
- At an international level there is no uniformity in the teacher preparation curriculum.

- The curriculum for initial and graduate teacher preparation is managed by government entities who do not take in account suggestions from in-service teachers.
- The Ministry of Education has the sole responsibility for preparing future teachers which undermines the competitive spirit and self-improvement motivation that should exist.
- Only courses taught by the Ministry of Education are considered in teacher salary increases.
- There are not enough graduate courses in Mathematics Education.
- There are insufficient opportunities for interchanges with other countries.
- There are not enough courses either online or face-to-face on the use of special Mathematics software.
- The scientific societies have little interaction with teachers.
- Many schools have inadequate electrical installations so many teachers do not use technological tools.
- The Internet is often unavailable in schools, particularly in rural areas.
- Classrooms are overcrowded.

Conclusions

Bolivian education merits greater and more detailed studies that will permit a reformulation of the curriculum in the Teacher Education Colleges (*ESFM*) and Academic Units (*UA*) for continuing and graduate education.

Various teacher associations have presented proposals that should be considered. One of the most important proposals is to consider learning situations that lead to a reflection on theory and practice thus seeking to enrich ideas and, as a consequence, working to improve classroom teaching. Such a suggestion challenges the current model of a repetitive and rote education that is predominant in schools.

The use of communication and information technologies in the classroom is currently minimal. These technologies are not being used adequately to maximize communication and learning of Mathematics.

The use of interactive programs and technological/multimedia resources such as dynamic software (Cabri II, GeoGebra), web pages and virtual platforms, is vital so that students can explore and describe relations and properties using visualization to manipulate shapes graphically and inductively.

It is fundamental to encourage attendance at seminars, conferences, workshops, congresses, and continuing education courses so that in the future teachers will have a global vision and be in permanent professional development.

References

Ministerio de Educación, Cultura y Deportes de Bolivia. (1999). *Diseño curricular base para la formación de maestros del nivel Primario.* La Paz: Viceministerio de Educación Inicial, Primaria y Secundaria.
Ministerio de Educación, Cultura y Deportes de Bolivia. (2000). *Diseño curricular base para la formación de maestros del nivel inicial.* La Paz: Viceministerio de Educación Inicial, Primaria y Secundaria.
Ministerio de Educación, Cultura y Deportes de Bolivia. (2002). *Proyecto inicial de Reforma de la Educación Superior.* Reforma Educativa, Viceministerio de Educación Superior, Ciencia y Tecnología (1era ed.). La Paz: Artes Gráficas Sagitario.
Ministerio de Educación, Cultura y Deportes de Bolivia. (2010a). *Diseño curricular base para la formación de maestros del nivel secundario.* La Paz: Viceministerio de Educación Inicial, Primaria y Secundaria.
Ministerio de Educación, Cultura y Deportes de Bolivia. (2010b). *Informes varios sobre la situación de la educación actual en Bolivia.* La Paz: Viceministerio de Educación Inicial, Primaria y Secundaria.
Ministerio de Educación, Cultura y Deportes de Bolivia. (2012). Boletin Informativo. La Paz: DGFM
Ministerio de Educación, Cultura y Deportes de Bolivia. (2015). Boletin Informativo. La Paz: DGFM
Ministerio de Educación, Cultura y Deportes de Bolivia. (2016). *Boletín Informativo.* La Paz: Universidad Pedagógica.
Ministerio de Educación, Cultura y Deportes de Bolivia. (2017). Boletin Informativo. La Paz: DGFM
Muñoz Reyes, J. (1980). *Geografía de Bolivia.* Academia Nacional de Ciencias de Bolivia. La Paz: Editorial Juventud.
República de Bolivia. (1994). *Compilación de leyes y otras disposiciones de la Reforma Educativa. Ley N° 1565.* La Paz: Editorial UPS.
República de Bolivia. (2006). Ley N° 070 de lo Educacion Avelino Siñani - Elizardo Perez. La Paz

Report on Mathematics Teacher Preparation in Ecuador

Margarita Martínez Jara, Paola Castillo Domenech, César Trelles Zambrano, Eulalia Calle Palomeque, Andrea Ayala Trujillo, Fredy Rivadeneira Loor and Roxana Auccahuallpa Fernández

M. Martínez Jara (✉)
Escuela Superior Politécnica del Litoral, Guayaquil, Ecuador
e-mail: mmartin@espol.edu.ec

P. Castillo Domenech · A. Ayala Trujillo
Universidad San Francisco de Quito, Quito, Ecuador
e-mail: acastillo@usfq.edu.ec

A. Ayala Trujillo
e-mail: aayala@usfq.edu.ec

C. Trelles Zambrano · E. Calle Palomeque
Universidad de Cuenca, Cuenca, Ecuador
e-mail: cesar.trellesz@ucuenca.edu.ec

E. Calle Palomeque
e-mail: eulalia.calle@ucuenca.edu.ec

F. Rivadeneira Loor
Universidad Técnica de Manabí, Portoviejo, Ecuador
e-mail: fredyrivadeneiraloor@gmail.com

R. Auccahuallpa Fernández
Universidad Nacional de Educación, Azogues, Ecuador
e-mail: roxaaf@gmail.com

Y. Y. Baldin and U. Malaspina (eds.), *Mathematics Teacher Education in the Andean Region and Paraguay*, https://doi.org/10.1007/978-3-319-97544-3_2

Introduction

Ecuador is a country situated on the northwest Pacific coast of South America. Currently, Ecuador has 24 provinces with 16 million inhabitants: 72% Mestizo, 7% Montubio,[1] 7% Afro-Ecuadorian, 7% Indigenous, 6% White, and 1% others. The area is 283,561 km^2, and the GDP per capita is over $4000. According to Web Datos Macro (n.d.) in 2012 the government of Ecuador invested €2,849,100,000 in education to serve a student population of 4,418,913; which is equivalent to $829 per student, much less than South Korea ($6732) or the United States ($10,995) during the same year (National Academies of Sciences, 2015).

In an effort to know the current initial and continuing mathematics teacher preparation situation, data from the Ministry of Education (MINEDUC), information from various universities in the country, and reference texts and documents available online have been examined. To reflect on the strengths, weaknesses and threats related to the teaching of Mathematics in Ecuador, the report summarizes not only education's historical context, but the main characteristics of the current structure for Mathematics teacher preparation.

Historical Summary

When formal education began, the Ecuadorean teacher was prepared as a generalist who taught all subjects in the same classroom. In 1863 there was a restructuring of public education in Ecuador. Education was to be offered in public establishments, as well as those run by religious orders, particularly Catholic. There were elementary and secondary schools, and universities. The La Salle Christian Brothers were in charge of elementary education, and the Jesuits of secondary and higher education. Free elementary education was established in 1871 and in 1899 the first Normal Schools for preparing teachers were founded. The Normal Schools were run by the Christian Brothers and nuns of the Sacred Heart. The creation of the Normal Schools facilitated not only a quantitative growth in education, but a qualitative growth as well (IESALC, 2004; OIM, 1994).

In 1895, President Eloy Alfaro, institutionalized secular public education and invested a great part of the State Budget in the change. At that time, the Mejía National Institute was created in Quito and Normal Schools were established in Guayaquil and Quito (Freire, 2008; MINEDUC, 2015b). In 1913 various German missions were contracted with the objective of developing a new curriculum for teacher preparation with a Herbartian focus.[2] As a result of the project with the Germans, 2400 teachers were prepared, 320 of whom where in Normal Schools dedicated to elementary education (MINEDUC, 2015b; UNESCO, 2004).

[1] Montubio are an indigenous mestizo people of coastal Ecuador. The Montubio were recognized by the government in the spring of 2001 after protests that included protracted hunger strikes. http://www.ijunoon.com/dictionary/Montubio/.

[2] Herbart indicated that interest is the cardinal concept of instruction, but is the end not the means.

The Faculty of Philosophy and Letters of the Central University of Quito was founded in 1928 and included an Institute of Pedagogy. The Institute was created to culminate all the efforts to improve elementary and secondary teaching. In 1930, an emphasis was given to rural education, taking into account particular rural characteristics. In 1950, 28 Normal Schools were created to increase teacher supply. Four years of study were required to obtain the title of Rural Teacher, while for Urban Teacher, it was six years (Freire, 2015).

In 1974, the Central University created a Bachelor's (*Licenciatura*) degree in Education Sciences that permitted secondary teaching at a higher salary. Thus, about 50% of the Normal Schools disappeared. Finally, in 2014 all the Normal Schools were eliminated and the universities took charge of all teacher preparation for elementary and secondary schools. Currently, there is an effort to organize the offering of programs to eliminate the excessive diversity of degrees and to bring more consistency to the basic curriculum. One example of this effort to organize is the proposal for generic curricula in areas such as education (CES, 2015).

Structural Aspects

From a regulatory point of view, the structure and function of the Ecuadorean educational system is presented in five normative documents: (1) the Constitution of the Republic of Ecuador (effective from October 20, 2008); (2) the Organic Law of Higher Education-*LOES* (published in the *Official Register* on October 12, 2010); and (3) its Regulations; (4) the Organic Law of Intercultural Education-*LOEI* (effective March 31, 2011); and (5) the Regulations of the Organic Law of Inter-cultural Education (codified and updated on August 25, 2014). Pre-university education is divided into three levels: Pre-Primary Education, General Basic Education-*EGB*, and General Unified Upper Secondary-*BGU* (*Bachillerato General Unificado*). Table 1 shows the current structure (Reglamento General LOEI, 2015).

Since 2015, compulsory, secular and free education begins at five years of age with Grade 1 of General Basic Education (the equivalent of Kindergarten) and ends with the third year of upper secondary (Table 2).

With respect to General Basic Education, the Ministry of Education has elaborated curricular guidelines for the main areas of study: language and literature, Mathematics, social studies and natural sciences. In the curricular guidelines, exit profiles for students finishing *EGB* and cross-cutting processes are established. The cross-cutting processes are interculturality, formation of a democratic citizenry, protection of the environment, health care, habits of recreation for all students and sex education for adolescents. Table 3 shows the integrative curricular processes[3] and learning processes for the content area of Mathematics.

---

[3]The "integrative curricular processes" are concepts of a great degree of generalization of the academic contents that articulate the whole curricular design in each content area, with a projection toward interdisciplinarity.

Table 1 Levels and sublevels of the Ecuadorean educational system in effect since 2010

Level	Sublevel	Characteristics	Ages
Pre-primary	Pre-primary 1	Not in school	Up to 3 years old
	Pre-primary 2	In school	From 3 to 5 years old
General basic education-*EGB*	Kindergarten	Grade 1[a] of *EGB*	From 5 years old
	Primary	Grades 2, 3 & 4 of *EGB*	From 6 to 8 years old
	Upper elementary	Grades 5, 6 & 7 of *EGB*	From 9 to 11 years old
	Lower secondary	Grades 8, 9 & 10 of *EGB*	From 12 to 14 years old
General unified upper Secondary-*BGU*		Upper secondary 1, 2 & 3	From 15 to 17 years old
Upper secondary in science			
Technical upper secondary			

[a]Ecuador classifies Kindergarten as Grade 1 so the rest of the grades are 1 higher than in most countries
Source Adapted from the Organic Law of Intercultural Education-*LOEI* (2015)

The Current Initial Preparation of Teachers of Mathematics

Current Structures for Initial Preparation

Ecuador does not currently have a clear structure that defines initial teacher preparation. In an effort to bring about some standardization in degrees and programs offered by higher education institutions, the current government has worked to establish generic curricula and the various education programs have redesigned their structures and curricula to be in harmony with the regulations issued by the Higher Education Council (*Consejo de Educación Superior-CES*). The universities are charged with preparing Mathematics teachers. Among the degrees that are offered are a Bachelor's (*Licenciatura*) in Education Sciences or in General Basic Education, and a Bachelor's in Experimental Sciences Pedagogy: Mathematics and Physics or a Bachelor's in Physics and Mathematics. Teachers who graduate in General Basic Education can teach Grades 1 to 7 (i.e. what used to be called primary or elementary in Ecuador and K-6 in many countries). Teachers who graduate with a degree in Experimental Sciences Pedagogy: Mathematics and Physics teach lower and upper secondary.

A representative sample of the curriculum for the Bachelor's in General Basic Education is that of the Pontifical Catholic University of Ecuador. In the case of the Bachelor's in Experimental Sciences Pedagogy, each university has a curriculum based on a proposal from the Mathematics Education Network. The structures and percentages of credits are summarized in Table 4.

Table 2 Hours of mathematics in elementary and secondary schools

	General basic education-*EGB*									Upper secondary			
Grade	G1	G2	G3	G4	G5	G6	G7	G8	G9	G10	HS1	HS2	HS3
Hours/week of mathematics	8	8	8	7	7	7	7	6	6	6	4	4	4

Source Adapted from MINEDUC (2015a, c)

Table 3 Processes for mathematics in general basic education and upper secondary

	General basic education	Upper secondary
Integrative curricular processes	"Develop logical and critical thinking to interpret and solve real life problems"	"Acquire mathematical concepts and tools that develop logical, mathematical and critical thinking to solve problems by using models"
Learning processes	Reasoning, proof, communication, connections and representation	Abstraction, generalization, conjecture and proof; integration of knowledge, communication of mathematical ideas; use of technology

Source Adapted from MINEDUC (2015a, c)

Table 4 Comparison of curricular structures

Bachelor's in general basic education		Bachelor's in experimental sciences pedagogy: Mathematics and physics	
Area	Weigh t (%)	Area	Weight (%)
Psychopedagogy	29	Professional Praxis Pedagogy (4% in Mathematics)	11
Theoretical and axiological foundations	18	Theoretical foundations (26% in Mathematics)	40
Instrumental content (2% in Mathematics)	17	Integration of knowledge, contexts and cultures (1% in Mathematics)	10
Student teaching (3% in Mathematics)	16	Student teaching (14% in Mathematics)	25
Graduation project	6	Epistemology, research methods and graduation project	10
English	14	Technology, communication and language (2% in Mathematics)	4

Source The authors

The curriculum for both programs is similar with respect to pedagogy, specialized and general courses, as well as student teaching and graduation projects. However, the distribution of courses has been into areas that do differ from one program to another. For General Basic Education there are six areas. The biggest component is psychopedagogy with 29%, while Mathematics reaches only about 5% spread across content and student teaching. This is due to the fact that preparation for this level is centered on pedagogy and methods for all subject areas. On the other hand, in the case of the Bachelor's in Experimental Sciences Pedagogy: Mathematics and Physics, 46% of the coursework is related to Mathematics with 28% dedicated to mathematical content and 18% to the Mathematics teaching methods.

Mechanisms for Recruiting Students

Strategies used by the institutions of higher education to attract students to their teacher education programs have changed since the implementation of the National Higher Education Examination (*Examen Nacional de Educación Superior-ENES*) and the National Leveling and Admissions System (*Sistema Nacional de Nivelación y Admisión-SNNA*).

According to SENESCYT (2012),[4] up until July of 2016, upper secondary school graduates seeking admissions to a public institution of higher education were required to take the *ENES*. The programs they are eligible for are determined by their scores. For admissions to Education and Medicine a score of a least 800 out of 1000 was stipulated. That requirement limited access to those programs in some of Ecuador's universities.

Currently, SENESCYT (2017) allows institutions of higher education to set their own minimum scores for admissions to programs in Education and Medicine by repealing what had been established as admissions regulations for the period 2012–2014.

A traditional way to recruit upper secondary graduates was to visit the main schools in a community to disseminate information on the importance of becoming a teacher. However, because of the difficulties of being admitted to the public institutions, many upper secondary graduates have opted to attend private universities. This has caused a great increase in the number of private institutions. They have not been prepared to meet the growing demand and have experienced difficulties in maintaining the personalized attention that they promote.

[4] SENESCYT is the National Secretariat of Higher Education, Science, Technology and Innovation.

Table 5 Undergraduate degrees of teachers of mathematics teachers

Undergraduate degree	Percent (%)
Bachelor's in Education Sciences: Mathematics and Physics	51
Engineering (Civil, Electrical, Systems)	13
Bachelor's in Pure Mathematics	4
Bachelor's in Education Sciences (Various majors: Art, Chemistry, Mechanics, Psychology)	18
Bachelor's in Business Administration	7
Other	7

Source The authors

Entities that Participate in the Preparation of Mathematics Teachers

Currently there is initial teacher preparation only at the universities. The government, through the *SNNA*, the *CES*, and the Council for Evaluation, Accreditation and Assurance of the Quality of Higher Education (*CEAACES*), has forcefully initiated a restructuring of the contents and methodologies of teacher preparation. There is no evidence of a significant impact by the Ecuadorean Society of Mathematics (*SEDEM*) as an advisor to the process. There is also not a clear line of external advice, although there have been talks and meetings with consultants from Finland, Spain, Belgium, Mexico and others.

Descriptive Statistics on the Teachers of Teachers

There, are tables that present data on some of the characteristics of the teachers who prepare future Mathematics teachers in Ecuador. Table 5 presents data on their undergraduate degrees while Table 6 considers graduate degrees. Table 7 has information on gender, nationality and type of employment within the institutions of higher education. Ranges of age and years of teacher experience appear in Table 8. These data are from only three of the universities with teacher preparation programs.

Institutional Relations Among the Actors and Entities

In Ecuador, at least 31 universities offer undergraduate or graduate programs related to Education. This report uses data that has been collected from universities in 12 of Ecuador's 24 provinces. These universities, independent of their locations, num-

Table 6 Graduate degrees of teachers of Mathematics teachers

Graduate Degree	Percent (%)
Master's in Education	38
Master's in Mathematics Teaching	18
Master's in Educational Projects	12
Doctorate in Educational Sciences (Not equivalent to a Ph.D.)	4
Master's in Information and Communication Technologies	4
Doctorate in Forest Sciences	2
Master's in Development of Thought	2
Higher Diploma in Pedagogy	2
Other Master's degree (Administration, Economics, etc.)	16
Without a graduate degree	2

Source The authors

Table 7 Gender, nationality and type of employment of teachers of teachers

Gender		Nationality		Type of Employment	
Female	31%	Ecuadorean	100%	Tenure track	58%
Male	69%	Other	0%	Adjunct	42%

Source The authors

Table 8 Age and years of teaching experience of teachers of teachers

Age		Years of teaching experience	
30–39 years old	20%	0–9 years	19%
40–49 years old	38%	10–19 years	28%
50–59 years old	33%	20–29 years	38%
60–69 years old	9%	30–39 years	16%

Source The authors

ber of Education programs or academic level, are representative of the universities that participate in teacher preparation. Only 48% (15 of 31) of the group offer programs related to Mathematics. Only one of the group is private. Only 8 of the 31 offer graduate programs in Education, and, of the 8, only 2 offer a specialization in Mathematics.

Each of the universities in Ecuador goes through a process of accreditation carried out by *CEAACES*. Since 2013, institutions of higher education have been classified using a set of methodologies for multicriteria analysis and the analysis of a cluster of data related to academic efficiency, research, structure and infrastructure (CEAASES, 2014). Based on the categories given by *CEAACES* two universities were classified as type A: the Polytechnic School of the Coast (*Escuela Politécnica del Litoral-ESPOL*) and the San Francisco University of Quito (*Universidad San Francisco de Quito-USFQ*). Twelve universities were classified in category B and the remaining

55% (17 of 31) were classified in categories C or D. Therefore, more than half of the universities that offer Education programs do not guarantee academic excellence, nor do they meet the goal of providing graduates with the necessary profile as elementary teachers or Mathematics teachers. Although the universities in Ecuador have been classified, the accreditation process at the program level is still in process. Very few programs have been accredited and none in Education or Mathematics Education.

Differentiation Between Elementary and Secondary

Historically, in Grades K-6 (now Grades 1–7 in Ecuador) there was one teacher per class of students that taught all subjects. In subsequent grades there are specialized teachers for each separate subject. Elementary and lower secondary teachers should have an undergraduate degree in Basic Education and those teaching in upper secondary should have an undergraduate degree in Education with a major in the disciplines they teach. With respect to Mathematics, that implies a major in Mathematics-Physics or something related. However, this does not occur in practice as can be seen in Table 9.

The Content for Teacher Preparation

Because the curricular structures, requirements and teaching methodologies for elementary and secondary are different, the analysis of the principle academic components of teacher preparation will be presented separately.

Academic Components for Mathematics Teacher Preparation Beginning with Lower Secondary School

The preparation of teachers to teach Mathematics at the upper secondary level is the responsibility of universities that have education programs with a major in Mathematics.

Each one of the programs offered by the 13 universities divides the academic structure into a theoretical part and a practical part. The academic components of the theoretical part can be classified into the following categories: Mathematics, Physics, General Pedagogy, Mathematics Pedagogy, and other (including administration, languages, technology and electives). To calculate the percentages represented by each of the categories an analysis was completed using the official programs of 11 of the 13 universities. Table 10 shows the averages for the academic components.

Table 9 Disciplines of teachers by grade level taught

	G1[a]	G2	G3	G4	G5	G6	G7	G8	G9	G10	G11	G12	G13	IB[b]
Mathematics (%)	0	0	0	0	0	0	0	2	4	3	4	4	3	5
Math and Education (%)	1	0	1	2	1	3	4	22	32	38	44	48	52	68
Education (%)	91	90	84	68	80	72	72	43	32	27	19	17	15	13
Other areas (%)	8	10	14	30	19	25	24	33	31	32	33	31	30	14

[a]G1 in Ecuador is K in many countries
[b]*IB* International Baccalaureate
Source MINEDUC (2015c)

Table 10 Average percentages of the academic components in the programs in Pedagogy of the Experimental Sciences: Mathematics and Physics at 11 universities

Mathematics	Physics	General pedagogy	Mathematics pedagogy	Graduation projects	Other
30%	17%	22%	4%	7%	20%

Source The authors

Table 11 Average percentages of the academic components in the programs in General Basic Education at 9 universities

Mathematics	Physics	General pedagogy	Mathematics pedagogy	Graduation projects	Other
5%	15%	44%	3%	12%	21%

Source The authors

The percentages found in the academic courses for future upper secondary teachers differ significantly from those for future elementary teachers. While the programs with a major in Mathematics and Physics have a near balance between content courses and pedagogical courses, the programs for future elementary teachers have mainly general pedagogy and psychology courses. The percentages of each of the academic components in the preparation of elementary teachers that is presented in Table 11 were calculated based on the programs at nine of the 15 universities that offer such programs.

The 21% found in the "Other" category, as is the case in Table 11, corresponds to courses in administration, languages and technology, as well as, electives. Just 5% is dedicated to Mathematics content courses and even less to courses on the teaching of Mathematics. These small percentages might explain the existing shortcoming among elementary teachers with respect to both mathematical content and pedagogy. Evidence of this shortcoming is presented in the report from INEVAL which indicates that 62% of teachers that take the "Be a Teacher" (*Ser Maestro*) test for professional advancement have insufficient mathematical knowledge by scoring no more than 547 points out of 1000, as is pointed out below in the section on "Strengths, Weaknesses, Opportunities and Threats."

Student Teaching

The current programs for future educators offered by Ecuadorean universities must include at least 400 h of student teaching which is about 7% of the total programs. In the new proposal, student teaching hours will be increased to 1800 which will be 25% of the total of 7200 h. As can be observed, the total number of hours will be increased, but the proportion dedicated to student teaching is not conserved and thus there is a challenge for implementation. The new proposal allows student teaching to be distributed throughout the programs beginning in year 1 as determined by each institution. This proposal, although approved by the Ministry of Education, has not been put into practice by the universities yet. All teacher education programs are currently undergoing a process of updating, adaptation and accreditation.

In Ecuador, at the present time, the student teaching component appears in the last semesters of the programs. That is, once a student has studied for three years (six semesters) they must complete student teaching during the fourth year. The length of programs and the distribution of courses depends on each university. Programs vary between eight and ten semesters, and on average student teaching begins in the fifth semester. At that time, students are required to put into practice all that they have learned in the university classroom and demonstrate that they have developed the specific skills and abilities that are necessary to be successful in their future career.

Student teaching varies according to the educational level for which they are preparing and according to the university offering the program. There are differences between student teaching for future elementary and secondary teachers. For future elementary teachers, student teaching has two phases. In the first phase, student teachers assist the regular classroom teaching in everything from planning to teaching of lessons. In the second, the student teachers plan and teach lessons on their own under the supervision of the classroom teacher.

In the case of secondary education, student teaching is a joint institutional process between the universities that prepare the future teachers and the schools that receive them. As part of this process the student teachers request the necessary permission from the schools to plan and teach certain classes. They are supervised by the classroom teacher who evaluates them using a form where they indicate the content knowledge of the student teacher as well as the techniques and methods used. The student teachers must include both evaluation and instruction activities. Thus, it is expected that the cooperating teacher will have a complete vision of the student teachers' abilities to teach.

Research on Mathematics Education and Academic Networks

In general terms, there are not clearly defined areas of research in Mathematics Education. Although some professors do educational research at their institutions, they are not directly involved in teacher preparation.

Universities Where Research Is Conducted

D'Amore (2006) in his *Teaching Mathematics* (*Didáctica de la Matemática*) seeks to connect Spanish-speaking Mathematics teachers with experts on Mathematics Education and researchers in the field of teaching Mathematics, given that research in Mathematics Education in Latin America is not particularly strong. Such is the case in Ecuador. In the 13 universities that offer programs with a major in teaching Mathematics or Sciences, only a few have created some momentum in the research field of Mathematics Education or have formally established research areas in Mathematics Education. Of the general list of universities with programs in Mathematics Education, the following have established research areas in the field:

- **The University of Cuenca**. Three main areas of research are being developed: (a) the social context of the actors in the Mathematics teaching and learning process; (b) methods, resources and evaluation centered on the Mathematics learning; and (c) information and communication technologies in the Mathematics learning.
- **The National Education University (*UNAE*)**. Research is being conducted in the Intercultural Bilingual Education program on the Ethnomathematics of the ancestral knowledge of native peoples.
- **The Central University of Ecuador**. Research is being conducted related to Ethnomathematics. In 2016, they hosted the "First International Congress on Ethnomathematics" and will the Second in October of 2017.
- **The Bolívar State University**. Areas of research in Mathematics Education have not been formally established. However, a formal research area in "Education and Knowledge" does exist. In their Education program there are four thematic components: (1) the diversity of learning-teaching, (2) educational correlations between the generation of knowledge and traditional techniques, (3) teaching and learning environments and strategies for socially and economically vulnerable groups, and (4) intercultural teaching in urban and rural contexts for alternative education.
- **The Private Technical University of Loja**. The strategic area is centered on relevant and high-level teaching, focusing on psychopedagogy, teaching methods and evaluation (UTPL, *s.f.*).
- **Simón Bolívar University—Quito**. In collaboration with the Santa Fe Institute, research is being developed on STEM education.

It is also important to point that some individuals who are not affiliated directly with a university do work in research on Mathematics Education. One example is Dr. Eduardo Molina who has been investigating the relationship between language and literature.

Research as a Component of Mathematics Teacher Preparation

Section 1.1.3 of a document published by the Council on Higher Education (CES 2015, p. 7) called "Generic Curriculum for Education Programs" states that "the pre-professional practice of action research is an organizing strand of the curriculum because it invigorates the necessary linkages between study and the encounter with the diversity of educational problems presented by the subjects, contexts, systems and phenomena." Therefore, in light of this paradigm, future teachers should earlier on be capable of conceptualizing research as a "guiding hand" in the educational process and in solving real life problems. This focus should favor contact with other disciplines and a vital connection between the model and reality. Nevertheless, so that these aspirations go beyond a simple lyrical enunciation, good instructor-researchers are needed to serve as models. There are few such models in Ecuador. Therefore, in a framework that included active participation of teachers in its development, the Ministry of Education's "10-Year Plan for Education 2016–2025" has been set in motion actions that should revitalize teaching as a career in Ecuador. It is hoped that many teachers will take advantage of opportunities for master's degree programs or graduate courses that have become available, and thus provide teacher-researchers in the field of Mathematics Education. Also, in 2017–2018, teachers will have the opportunity to take part in an international master's degree in Secondary Education with a major in Mathematics that is being offered through a collaboration between Ecuador's National Education University (UNAE) and the University of Barcelona. The program will lead to the awarding of degrees from both institutions. A goal of the program is to prepare teacher-researchers, even though there is not much time available to teachers outside their work schedules.

The Role of Networks and/or Academic Communities in the Generation of Mathematics Teacher Preparation Policies

In Ecuador, the Ecuadorean Mathematics Society (*SEDEM*) plays an important role in disseminating research in Mathematics. *SEDEM* is a non-profit, non-governmental scientific association. It was formed in 1967 by individuals working in Mathematics. The society establishes norms and regulations for the publication of research. It is also responsible for the organization of the Mathematics Olympiads for children and youth. However, it has not yet generated policies related to the preparation of teachers, nor participated in the development of school-related legislation (SEDEM, 2008).

According to Montalvo, Ormaza, and Espinosa (2015) in the Department of Research of the Ministry of Education, no connection exists between educational researchers and educational policies of the Ministry of Education. The Ministry does not have defined research areas, nor does it keep an inventory of research that has been conducted at the national level. This explains why research has not permeated the reality of schools.

Upon searching in the Scopus, Springer, Ebsco, Dialnet and Scielo data bases it can be determined that only 15 articles have been published about Ecuador. One was published in 2011, and the rest were distributed among 2015, 2016 and 2017. The universities associated with those 15 articles are the University of Cuenca, the Polytechnic School of the Coast, the National Polytechnic School, the Central University of Ecuador, The Equatorial Technical University, the University of Guayaquil, the University "Luis Vargas Torres," the Secular University "Eloy Alfaro," and the Ambato Technical University. The main themes covered in the mentioned articles are theoretical reflections on the process of teaching and learning Mathematics, Ethnomathematics, curricular proposals, student mathematical abilities, and the use of technological resources and learning styles. The articles report research results on various educational levels, but the most common are preprimary, elementary and higher education. The limited research literature generated in the country suggests that there is an urgent need to coordinate the various research initiatives that currently are often working in isolation.

Continuing Preparation

Currently there is not a defined structure for teacher continuing preparation. The universities have been offering courses, workshops, certificate programs and even Master's degrees. Given the lack of content knowledge demonstrated in the tests given by INEVAL, the government, through the Ministry of Education, has promoted and financed courses and Master's degrees. Since 2008 continuing preparation course have been organized on themes related to the *SER* test (MINEDUC, 2013a). However, programs have not been offered that address the need for more researchers in Mathematics Education.

Entities in Charge of K-12 Continuing Teacher Preparation

The current government has selected type A and B universities to offer courses for continuous teacher preparation in various areas, including Mathematics. Additionally, actions have been implemented with the goal of improving the way Mathematics is taught in Ecuador. Among the actions are those that involve.

Leadership from the Ministry of Education:

- The *SíProfe* Project was part of the 10-Year Plan for Education 2006–2015, in which the society and the government of Ecuador were involved with respect to quality of the education system and the performance of teachers (Portaluppi, 2012). During 2010, SíProfe offered 18 professional development modules with the participation of 271,681 teachers. Among the courses offered were Critical Reading, Teaching Elementary Mathematics, Pedagogy and Teaching, Introduction to Information and Communication Technologies, and updates on the elementary curriculum in various subjects and grades (MINEDUC, 2011).
- Initial Teacher Preparation: an extension of *SíProfe*.
- Master's Program with Spanish Universities: Beginning in May of 2014 the Ministry of Education in collaboration with the University of Barcelona, the Autonomous University of Madrid, the Complutense University of Madrid, and the National Distance Education University began various Master's programs that

have benefited approximately 2400 teachers (MINEDUC, 2015d; Presidencia de la República del Ecuador, 2016).

Leadership from *SENESCYT* through *SNNA*:

- Specialization in Teaching Mathematics
- Preparation of Mathematic Mentors.

Methodologies Used in Continuing Teacher Preparation

According to the Ministry of Education (MINEDUC, 2014a) the purpose of teacher preparation is to provide teachers with the tools to develop specific knowledge thus preparing professionals capable of teaching, generating and transmitting the necessary knowledge and values. Therefore, the methodology of the courses that the Ministry of Education offers must have the following characteristics:

- It is necessary to reinforce the stage of Conceptual development because there has been a tendency to cover material without consolidating the foundations that are necessary to support the learning spiral (Chacón & Valarezo, 2011).
- There must be self-directed and group work and readings, or collaborative activities.
- Active participation of the teachers in reflections and whole group presentations to permit an appreciation of diverse criteria and ways of transmitting mathematical knowledge.
- The strengthening of the practice of teaching by having the course participants develop sessions where they demonstrate the learning cycle.
- Information and communication technologies should be incorporated into the learning process.

The majority of the participating teachers are from the public system. Recently participation has often been mandatory. Also, teachers who fail the courses have to pay for them. Nevertheless, given that only a few universities and institutes have been authorized to give the courses, there are problems for some teachers to travel to the course sites and also the problem of the great diversity of academic backgrounds among the participants.

Recent Actions in Initial and Continuing Teacher Preparation

With respect to initial teacher preparation, *SENESACYT* set, starting from 2012, a minimum of 800 out of 1000 points on the *ENES* test to be eligible for admissions to teacher preparation programs. That was an effort to attract the best prepared high school graduates to become teachers and thus achieve a high level of university education, particularly in Mathematics Education. The strategy was unsuccessful.

Table 12 Number of teachers receiving professional development in Mathematics from *SíProfe* 2010–2013

	2010	2011	2012	2013
Updating Elementary Curriculum	21,259	7273	619	0
Updating Lower Secondary Curriculum	12,215	3261	590	783
Teaching Mathematics	16,648	9215	676	3729

Source MINEDUC (2014b)

Between 2012 and 2016 no one applied to become a Mathematics teacher. Therefore, in the new admissions regulation the 800 points requirement has been rescinded and it is hoped that more students will apply.

Also, the University of Cuenca (*UC*), the Central University of Ecuador (*UCE*), the Technical University of the North (*UTN*), the National University of Chimborazo (*UNACH*) and the Technical University of Manabi (*UTM*) have formed a network that is in the process of redesigning the teacher preparation curriculum, incorporating courses on the Teaching of Mathematics and Physics, Software for Teaching Mathematics and Physics, Ethnomathematics, Discrete Mathematics and Financial Mathematics. These new courses are in response to the current needs of society.

Since 2007 the Ministry of Education has been designing programs for in-service teacher development. One of these programs was the Integral System of Educational Professional Development (*Sistema Integral de Desarrollo Profesional Educativo-SíProfe*). *SíProfe* offered courses related to school Mathematics from 2010 to 2013:

- Updating and Strengthening the Elementary School Mathematics Curriculum, 20 h.
- Updating and Strengthening the Lower Secondary School Mathematics Curriculum, 20 h.
- Teaching Mathematics, 40 h.

Table 12 shows the number of teachers who received profession development from *SíProfe* during the period from 2010 to 2013.

The goal of the program was to provide Mathematics professional development for Elementary and Lower Secondary teachers. However, teachers from other disciplines took advantage of the courses. The proposed goal was not completely met for four fundamental reasons:

1. Not all the Elementary and Lower Secondary teachers of Mathematics could participate in the professional development programs.
2. The programs were of relatively short duration. For example, a 40-h course on the Teaching of Mathematics is insufficient given the difficulties of the subject.

Table 13 Percentage of teachers receiving professional development up to 2015

	Preprimary and Kinder (%)	Elementary (%)	Lower and Upper Secondary Mathematics (%)
2014	8	15	19
2015	68	45	41
Pending	25	40	40

Source MINEDUC (2013a)

3. The physical environment for the courses was not always conducive to adult learning as often the seating was for children.
4. The exit evaluation for the Teaching of Mathematics course was not aligned with the contents of the course and many teachers failed.

Beginning in 2014, the name for *SíProfe* was changed to Teacher Preparation. The National Directorate for the Continuous Preparation of the Ministry of Education in agreements with category A and B institutions of higher education, and with organizations like the National Institute of Higher Studies and the Organization of Ibero-American States via the Center for Higher University Studies, have offered professional development courses intended to bring improved attention to all the areas of identified needs for Ecuador's 210,850 teachers. Table 13 presents a summary of the percentages covered up to 2015 (MINEDUC, 2013b).

Additionally, in May of 2014, the Ministry of Education initiated in an agreement with Spanish universities the first phase of Master's degree programs that have benefitted more than 2400 teachers. Thus, for example, the University of Distance Education of Spain (*UNED*) admitted 960 teachers in four sites: Quito, Cuenca, Loja and Guayaquil. By December of 2015 approximately 240 teachers had graduated. In 2015, prestigious Ibero-American universities offered 10,000 online places for Master's and Specializations.

Future of Initial and Continuing Teacher Preparation

To provide answers concerning the future of teacher preparation in Ecuador the opinions of a group of Mathematics teachers at various levels were sought. The following two points summarize their aspirations:

- That with a generational change in the teachers and the paradigms, innovative and creative methodological strategies will be used so that students will be critical, reflective and truly develop the mathematical aptitudes that our society demands.
- That the majority of teachers in Ecuador will be able to develop their professional work to meet the minimums that have been established in the standards for educational quality (MINEDUC, 2013b).

Government authorities reiterate in their discourse a recognition of the role that education plays in scientific and technological development. However, given the immediate political rewards from investments in infrastructure, it is highly probable that infrastructure will continue to receive the largest budgetary contribution. This is a real paradox. On the Student *SER* evaluation carried by *INEVAL* in (2013) it was shown that public schools with smaller budgets actually performed better than private schools and schools that receive some public funds. Those results corroborate earlier Ecuadorean studies (Fundación Ecuador. Contrato Social por la Educación y Grupo FARO, 2006) as well as international studies (Compañía auditora McKinsey, 2007) that established that it is the quality of the teachers that determines the academic success of students.

Strengths, Weaknesses, Threats, Opportunities and Challenges

Strengths

- The existence of an appreciable number of young motivated teachers who have the energy necessary to continue with professional development and updating their knowledge.
- Academics in the universities have the solid base of experience needed to strengthen secondary teachers.

Weaknesses

- The lack of teachers specialized in Mathematics and a scarcity of official data on the demand for the next few years. Table 9 above shows that the majority of teachers of Mathematics have degrees in Education with a curriculum that is weak with respect to mathematical concepts.
- The decrease in the proportion of female teachers is notable as the grade level increases (see Table 14). The same pattern is repeated in regional areas. This has an effect on the promotion of careers in science and technology among female students.
- Teachers receive relatively little social recognition and low salaries.
- The lack of student interest in developing skills in Mathematics, caused by the mechanistic and algorithmic ways that classes are taught.
- Little participation by Mathematicians in the professional development workshops offered by the university so that mathematical concepts can be emphasized.
- Frequently there is no followed up and continuing support nor evaluation of the impact of workshops. The workshops do not respond to the real needs and moti-

Table 14 Percentage of female and male teachers in Mathematics Education

	G1	G2	G3	G4	G5	G6	G7	G8	G9	G10	G11	G12	G13	IB[a]
Female (%)	87	78	71	67	61	49	50	48	42	37	35	29	28	30
Male (%)	13	22	29	33	39	51	50	52	58	63	65	71	72	70

[a]*IB* International Baccalaureate
Source MINEDUC (2015d)

Table 15 Percentage of teachers of Mathematics as classified by depth of content knowledge by the *"Be a Teacher" (Ser Maestro)* evaluation of all public school mathematics teachers in 2016

Beginning	Basic	Favorable	Excellent
19.3%	70.6%	9.9%	0.1%

Source INEVAL (2013)

Table 16 Percentage of teachers seeking recategorization in 2014 as classified by "quality of achievement" in Mathematics

Insufficient	Good	Very good	Excellent
62%	25%	11%	2%

Source INEVAL (2014)

vations of the teachers. Instead they reflect the criteria of personnel working at the central level.

- There is a lack of adequate infrastructure, physical and virtual libraries, that respond to the requirements of the 21st century.
- In Ecuador there is not adequate access to Master's and doctoral degrees in Mathematics Education.
- There is an absence of research in Mathematics Education and little integration into collaborative networks.
- There is a lack of a generalized and strengthened curriculum for the Bachelor's in Education Sciences with a major in Physics and Mathematics.
- The insufficient depth of cognitive knowledge among a significant percentage of Mathematics teachers as is shown in the publication of results of testing by *INEVAL*. The average score in Mathematics on the *"Be a Teacher" (Ser Maestro)* evaluation used for teachers seeking recategorization in 2015 was 547 out of 1000 (INEVAL, 2013) (See Tables 15 and 16 for more results).

Threats

- The increase in the hours of student teaching from 400 to 1800 in teacher preparation programs has presented challenges to the universities. In some cases, they are eliminating important courses. In others, courses are being created that are called student teaching, but are not really developing practical classroom skills.
- Continual changes in the Organic Law of Higher Education (*LOES*) and its related Regulations generate uncertainty as they often establish new working days and work demands.
- Lack of communication among the main actors involved in education and, in particular, in Mathematics Education.

Opportunities

- The tendency on the part of the government and communication media to value the different areas of knowledge, including scientific thought.
- Technical and technological development is motivating students and teachers to have and demand better preparation in this area.
- The recognition that education is a priority area in the plan of the government with the consequent allocation of a greater budget to teacher professional development.
- Access for teachers to graduate programs offered by top national and international universities by teachers on the coast and in the mountains.
- Curricular redesign with an integrated vision of theory and practice. The challenge is to reconstruct the curriculum so that the increase in the hours of student teaching is the result of an analysis and selection of courses so that there can be an implementation with theory and practice as one unit.
- The universities in Ecuador are creating graduate programs that strengthen the sciences in the country and teachers will have access to those programs.

Challenges

- Attract and recruit enough high school graduate to meet the demand for teachers as current teachers retire and the school population grows.
- There is a need to prepare future teachers to support female students, particularly given the predominance of male teachers in the upper grades, as shown in Table 14.
- It is important to have a campaign to revitalize the role of Mathematics teachers in the scientific and technological development of the country.
- Prepare enthusiastic teachers that love Mathematics, and teach to value and develop it. This implies overcoming the limited demand for the programs in Mathematics and Physics that currently exist, and increase the social recognition and salaries in the profession.
- Teacher professional development that really provide the additional needed preparation.
- As is seen in the data from the Ministry of Education in Table 17 a very small percentage of teachers have Master's degrees.
- Educational authorities must improve the mathematical quality of in-service workshops by encouraging the participation of Mathematicians and must increase the opportunities for Master's and Doctoral level study in Mathematics and Physics Education.
- Collaborative work among the State, the academy and teachers in order that educational policies are not affected by political changes.
- Advantage should be taken of the goals of the National Plan for Well-Being (*Plan Nacional del Buen Vivir*) that seek to "improve the quality of education in all of its levels and forms…" (CES, 2015, p. 2).

Table 17 Relative frequency of highest degrees of teacher by grades in which they work

	G1	G2	G3	G4	G5	G6	G7	G8	G9	G10	G11	G12	G12	IB[a]
Less than a Bachelor's (%)	17	17	19	40	26	21	23	8	6	4	3	3	3	4
Bachelor's (%)	74	63	63	54	63	73	67	78	76	75	74	72	67	50
Specialization (%)	2	2	2	2	3	2	4	4	6	6	6	6	7	5
Master's or higher (%)	7	6	6	4	8	4	6	10	12	15	17	18	21	41

[a] *IB* International Baccalaureate
Source MINEDUC (2015d)

Conclusions

The history of reforms in Mathematics Education show a deep and integral failure to change the structure of Mathematics teacher preparation in Ecuador. On the other hand, for decades teacher unrest has been externalized through permanent conflict and the struggle for better salaries. A consequence has been the erosion of the social prestige of the teaching profession. Evaluation processes have shown the low academic level of public school teachers (INEVAL, 2013).

Critics of the structure of initial teacher preparation point out that the Mathematics curriculum is too focused on contents that are little related to reality. This is reinforced by presenting each subject independently, each one disarticulated from the others, which does not permit interdisciplinary connections. Each course also has too much material to be covered and few opportunities for the necessary sense making.

There is little educational research and what does exist is disconnected from classroom practice. It is important to recognize that the current government of Ecuador has presented initiatives and promoted policies to face educational problems. Some of the initiatives are relevant, such as increasing the budget for education, developing infrastructure, an effort at restructuring careers in education, offering undergraduate and graduate courses for teacher development. However, they are not adequately structured to meet the complexities of the situation. An integral, systemic and radical vision is required. The vicious cycle of teachers who do not enjoy Mathematics transmitting that dislike to the new generations must be stopped. It is important to have teachers who are passionate for discovery, experimentation and success in order to motivate the young, and enhance scientific and technological development. A search for reasons must be reinserted into classrooms, instead of limiting classes to empty, mechanical and boring algorithmic repetitions. We endorse one of the conclusions of the Social Contract for Education that proposes that "a social and conscious reaffirmation of the value of teachers of Mathematics, and attraction and cultivation of the best talents of the new generations are urgent goals" (Fabara, 2013, p. 12). Nevertheless, all of this need to be framed in a consolidated State policy that is independent of the government in power, and that provides a flexible structure, priority to the development of critical thinking and continuity to the efforts to transform Mathematics Education in Ecuador.

References

CEAACES. (2014). Categorización de universidades. Recuperado de http://www.ceaaces.gob.ec/sitio/categorizacion-de-universidades/.

CES. (2015). Currículo Genérico de las Carreras de Educación. Comisión Ocasional de Educación. Recuperado de http://www.ces.gob.ec/doc/Talleres_Carrera_de_Educacion/Curriculo_Generico/currculo%20genricodelascarrerasdeeducacin.pdf.

Chacón, M., & Valarezo, M. (2011). Didáctica de las Matemáticas. MINEDUC.

Compañía auditora McKinsey. (2007). "Cómo hicieron los sistemas educativos con mejor desempeño del mundo para alcanzar sus objetivos". Recuperado de http:// www.oei.es/pdfs/documento_preal41.pdf.
D'Amore, B. (2006). *Didáctica de la Matemática*. Colombia: Magisterio.
Fundación Ecuador. Contrato Social por la Educación y Grupo FARO, ISBN-9978-45-344-X. (2006). Calidad con Equidad: El desafío de la educación ecuatoriana: Informe de Progreso Educativo Ecuador 2006, Pag 8.
Fabara, E. (2013). Cuadernos del Contrato Social por la Educación. Cuaderno 8: Estado del Arte de la Formación Docente en el Ecuador. *8*(1), 97–98. Recuperado de http://campanaderechoeducacion.org/fresce/wp-apyus/wp-content/uploads/2015/09/O69-IV-ECU-A2-Pu-Inv.1.pdf.
Freire, A. E. (2008). Medición de la calidad de la educación fiscal en el oeste de Guayaquil, a través de los niveles de estudiantes de décimo año de educación básica. Recuperada de https://www.dspace.espol.edu.ec/bitstream/123456789/19302/1/CAPITULO1.pdf.
Freire, C. (2015). Hitos de la historia de la educación en el Ecuador, siglos VXI – XX. Para el aula. *13*(4). Recuperado de http://www.usfq.edu.ec/publicaciones/para_el_aula/Documents/para_el_aula_13/pea_013_0004.pdf.
IESALC. (2004). Educación Superior Virtual en América Latina y el Caribe. Recuperado de file:///C:/Users/MONICA/Downloads/EducVirtual_ALC.pdf.
INEVAL (2013). Ineval presenta los resultados de la evaluación Ser Estudiante 2013. http://www.evaluacion.gob.ec/ineval-presenta-los-resultados-de-la-evaluacion-ser-estudiante-2013/.
INEVAL (2014). Ser maestro recategorización. Recuperado de http://www.evaluacion.gob.ec/resultados/SM-como-nos-fue.
MINEDUC. (2011). Lineamientos pedagógicos SiPROFE.
MINEDUC. (2013a). Indicadores educativos 2011–2012. Recuperado de http://educacion.gob.ec/wp-content/uploads/downloads/2013/10/Indicadores_Educativos_10-2013_DNAIE.pdf.
MINEDUC. (2013b). Porcentaje de docentes a capacitar el 2015. Recuperado de http://educacion.gob.ec/formacion-conitnua-docente/.
MINEDUC. (2014a). El Ministerio de Educación invita a participar en las preinscripciones para los cursos de "Formación Continua. Recuperado de http://educacion.gob.ec/el-ministerio-de-educacion-invita-a-participar-en-las-inscripciones-para-los-cursos-de-formacion-continua/.
MINEDUC. (2014b). Estadísticas SíProfe. Recuperado de http://sime.educacion.gob.ec/Modulo/SIPROFE/index.php?mp=9_0.
MINEDUC. (2015a). Asignaturas del primer año del bachillerato unificado. Recuperado de http://educacion.gob.ec/malla-curricular-bachillerato-general-unificado/.
MINEDUC (2015b). Base legal. Recuperado de http://www.educacion.gob.ec/wp-content/uploads/downloads/2012/09/A1_Base_Legal_11.pdf.
MINEDUC. (2015c). Malla curricular de educación general básica. Recuperado de http://educacion.gob.ec/malla-curricular-educacion-general-basica/.
MINEDUC. (2015d). Base de datos de carrera profesional educativa. MINEDUC-AC-2015-09585.
Montalvo, N., Ormaza, J., & Espinosa, F. (2015). Entrevista de M. Martínez [Cinta de audio], Departamento de Investigación del Ministerio de Educación, Guayaquil.
National Academies of Sciences, Engineering, and Medicine (2015). Mathematics curriculum, teacher professionalism, and supporting policies in Korea and the United States. Summary of a workshop. Washington, DC: National Academies Press.
OIM. (1994). Informe OIM-Ministerio. Recuperado de http://www.oei.es/quipu/ecuador/#sis.
Portaluppi Castro, C. (2012). El Sistema Integral de Desarrollo Profesional Educativo (SíPROFE): Una mirada desde los involucrados en el sistema. Recuperado de http://dspace.casagrande.edu.ec:8080/bitstream/ucasagrande/386/1/Tesis401PORs.pdf.
Presidencia de la República del Ecuador. (2016). Noticias. Recuperado de http://www.presidencia.gob.ec/2322-docentes-del-magisterio-fiscal-culminaron-maestrias-en-universidades-espanolas/.
Reglamento General a la LOEI. (2015). Reglamento general a la LOEI. Recuperado de http://educacion.gob.ec/wp-content/uploads/downloads/2016/03/LOEI-Actualizado.pdf.

SEDEM. (2008). Estatutos vigentes. Recuperado de http://www.ceaaces.gob.ec/sitio/categorizacion-de-universidades/.
SENESCYT. (2012) (21 de noviembre de). Acuerdo 2012– 076. Reglamento del sistema nacional de nivelación y admisión. Recuperado de http://www.unemi.edu.ec/images/pdf/Reglamentosinternos/Reglamento-del-Sistema-Nacional-de-Nivelacion-y-Admision.pdf.
SENESCYT (2017) (20 febrero 2017) Reglamento del sistema national de nivelaciony admision. Recuperado de. https://www.unemi.edu.ec/images/pdf/Reglamentosinternos/Reglamento-del-Sistema-Nacional-de-Nivelacion-y-Admision.pdf
UCuenca. (s.f.). Universidad de Cuenca. Recuperado de http://www.ucuenca.edu.ec/.
UNESCO. (2004). Procesos de Formación de los Docentes por parte de Universidades e institutos pedagógicos del Ecuador. Recuperado de http://unesdoc.unesco.org/images/0014/001404/140489s.pdf.
UTPL. (s.f.). Universidad Particular de Loja. Recuperado de http://www.utpl.edu.ec/.
Web Datos Macro. (s.f.) Expansión gasto público educación. Recuperado de http://www.datosmacro.com/estado/gasto/educacion/ecuador.

Report on Mathematics Teacher Preparation in Paraguay

Gabriela Gómez Pasquali, Diana Giménez de Von Lücken, Osvaldo Ramón Vega Gamarra, Margarita Sanabria, María Clara Santa Cruz Cosp, Elisa Maidana de Giangreco, Jorge Daniel Mello Román and Richard Solís Argüello

G. Gómez Pasquali (✉) · M. C. Santa Cruz Cosp
OMAPA (Organización Multidisciplinaria de Apoyo a Profesores Y Alumnos), Asunción, Paraguay
e-mail: gabriela@omapa.org.py

M. C. Santa Cruz Cosp
e-mail: claritasanta@gmail.com

D. Giménez de Von Lücken
Universidad Católica Nuestra Señora de La Asunción, Asunción, Paraguay
e-mail: dianagimenez68@gmail.com

O. R. Vega Gamarra
Universidad Nacional de Asunción, Asunción, Paraguay
e-mail: vega.gamarra@gmail.com

M. Sanabria
Juntos Por La Educación, Asunción, Paraguay
e-mail: margarita.sanabria@juntosporlaeducacion.org.py

E. Maidana de Giangreco
Instituto Superior de Educación Dr. Raúl Peña, Asunción, Paraguay
e-mail: mami447@hotmail.com

J. D. Mello Román · R. Solís Argüello
Universidad Nacional de Concepción, Concepción, Paraguay
e-mail: jorgedanielmello@hotmail.com

R. Solís Argüello
e-mail: richardrsa1985@gmail.com

Y. Y. Baldin and U. Malaspina (eds.), *Mathematics Teacher Education in the Andean Region and Paraguay*, https://doi.org/10.1007/978-3-319-97544-3_3

Introduction

After 35 years of dictatorship, the 1992 Paraguayan Constitution defined Paraguay, a country located in the middle of South America, as a democratic, non-denominational and unitary State under law. Paraguay was declared a bilingual nation, establishing Spanish and Guaraní (spoken by 80% of the population) as the official languages. It is also recognized in this Magna Carta that everyone has the right to an integral and permanent education.

All human beings have a right to an education that permits them to acquire knowledge and live a full life. Although Law N° 1.264 of May 14, 1998 extends this constitutional principle and grants to education a priority in the fight against inequality and the opening of new opportunities to all the inhabitants of the country, in Paraguay less than 4% of Gross Domestic Product (GDP) is invested in education. And this, despite the fact that, in the last five years the country is experiencing unprecedented economic growth. This limited investment places Paraguay among the countries of the region that invest the least in education.

It is a concern that in Paraguay, a country of almost 7 million inhabitants with over 50% under 24 years of age, the quality of education is very low, even though there has been an increase in enrollment. The Inter-American Development Bank (BID, 2012) in its report on critical thinking skills in elementary education in Paraguay recognizes the limitations and low achievement of students in Mathematics and Sciences. The conclusion is that a great number of students do not acquire sufficient skills and knowledge to function effectively in society. That conclusion can also be seen reflected in UNESCO's (2014) Third Regional Comparative and Explanatory Study (TERCE) on Mathematics. Paraguay was one of three countries that showed no improvement in Grade 3 and one of 2 countries that showed no improvement in Grade 6.

Moreover, in the Global Competitiveness Index (GCI) prepared by the World Economic Forum (WEF) for (2013)–2014, out of a ranking of 148 countries Paraguay was 138 in the quality of its education system, 142 in relation to the teaching of Mathematics and Sciences, and 134 with respect to the availability of training and research. In a study on critical thinking carried out for the BID (2012) it was concluded that Mathematics teaching depends excessively on memorization of content and the repetition of mathematical procedures, combined with exercises done by rote. Without a doubt, one of the factors that determines such low results is the deficient quality in the initial preparation of teachers and the lack of continuing professional development.

This report outlines a series of findings and analysis on initial and continuing preparation of teachers and other professionals in Mathematics Education in Paraguay. It has as its objective to contribute to and collaborate with the Latin American community in efforts to provide answers to the educational problems and challenges in the region, particularly those related to the quality of education.

In the first part of the document, historical antecedents of teacher preparation, the current structure of the Paraguayan education system, as well as, the current situation with respect to initial and continuing teacher education will be described. Then, some of the more recent actions related to initial and continuing teacher preparation, as well as, research in Mathematics Education and its relation to teacher preparation will be presented. Finally, some of the challenges and opportunities for teacher preparation in Paraguay will be considered.

Teacher Preparation in Its Historical Context

From the Colonial Period to the 19th Century

During the Hispanic colonial period in Paraguay, education was not among the priorities of the Spanish authorities. Nevertheless, city leaders in Asunción always showed an interest in education. Both elementary and secondary schools were opened. It is important to mention that in the 16th and 17th centuries religious orders were the main promoters of Culture ([OEI-MEC], 1994a).

The first elementary schools, opened in Asunción in 1542 and 1543, were no more than places for religious indoctrination and for training in local trades (called *casas de doctrina* in Spanish). The first mestizos and local indigenous youths attended them. Reading, writing and arithmetic were taught when the mestizos began to demonstrate qualities and aptitudes that motivated their Spanish progenitors to offer such learning (Durán, 2012).

As the colonial population expanded toward rural areas, the possibilities for formal education were greatly affected by a scarcity of teachers. The situation was even worse for the indigenous population. The three centuries of colonial rule left as a result a very inadequate education system ([OEI, MEC] 1994a).

Beginning with independence from Spanish domination in May of 1811, education became a focus of attention for successive governments. The first teacher preparation institutes were established for the primary level in 1812 and the Paraguayan Normal School opened in 1870 ([OEI, MEC], 1994a).

From the 20th Century to the Present

The institutional basis of teacher preparation in Paraguay is similar to that in many Latin American countries. The origins are in the Normal Schools, created in the 19th century, and were at what is now considered to be the upper secondary level. In the 1970s, the Normal Schools became institutions of higher education. Thus, the years in school required to enter teacher preparation was increased, as well as the number of years of preparation needed to obtain the corresponding certification (Vaillant, 2009).

Below the main teacher preparation milestones in the last few decades will be detailed.

In 1968, Executive Order N° 31.003/1968 created the Higher Institute of Education (*Instituto Superior de Educación-ISE*) to address the need for qualified teaching personnel, develop those that did not have adequate pedagogical preparation, and implement curricular reforms in teacher preparation (PEP 1968). The purpose of the ISE was to be the governing body to oversee the implementation of policies related to teachers promulgated by the Ministry of Education. It continues to be the main reference point for initial teacher preparation in the country.

Later, the Plan for Educational Development (1969–1980) was put in place. Its main goal was to significantly improve Paraguayan education. To do so, it renovated the teacher preparation curriculum, aiming at modernizing teaching and educational administration (UNESCO, 1985).

From 1970 to 1974, with the closing of many Normal Schools, a process of proliferation of Teacher Preparation Institutes (Institutos de Formación Docente-*IFD*) for future preprimary, elementary and secondary teachers began ([OEI-MEC] 1994b).

In 1973, a new reform called "educational innovations" was implemented. It redefined the objectives and goals of Paraguayan education following the guidelines of the dictatorship (Zayas Rossi, 2015). At that time, the government had an active role in educational policy and in the processes of educational reform that were being introduced. The Stroessner dictatorship exercised control over what teachers could teach, with an emphasis on rote learning without questioning. According to Zayas Rossi (2015), the teachers that were used by the dictatorial regime kept quiet about flaws in the curriculum and in the preparation of teachers.

In 1998, with the passage of General Law of Education (Law 1264/1998) (CNP, 1998), the universities and higher institutes, among them the *ISE*, were granted the authority to offer Bachelor degrees (*Licenciaturas*) in the Education Sciences. This law also established the autonomy of the universities and higher institutes.

Expansion of the Teacher Preparation Institutes (IFD)

During the educational reforms implemented from 1993 to 1998, compulsory schooling was raised to Grade 9 (Law 1264–1998) (CNP, 1998). This implied a massive increase in school enrollment and the resulting demand for an increased number of teachers. As result of that demand, new Teacher Preparation Institutes (*Institutos de Formación Docente-IFD*) were opened. Demelene (2007, cited in Banco Mundial 2013) in referring to the expansion of the *IFD* states:

> In 1990, Paraguay had 17 *IFD* (16 public and one private). Fully 86% of the current *IFD* were created after the beginning of the educational reform of the mid-1990s. The growth has been concentrated in the private sector, which today accounts for approximately 65% of the *IFD*. With the growth in the number of *IFD* there was a corresponding explosive growth from 1995 to 2000 in the number of students being prepared to be teachers. That growth, in enrollments as well as number of institutions, besides having produced a lower quality in teacher preparation, generated an important problem of oversupply of teachers.

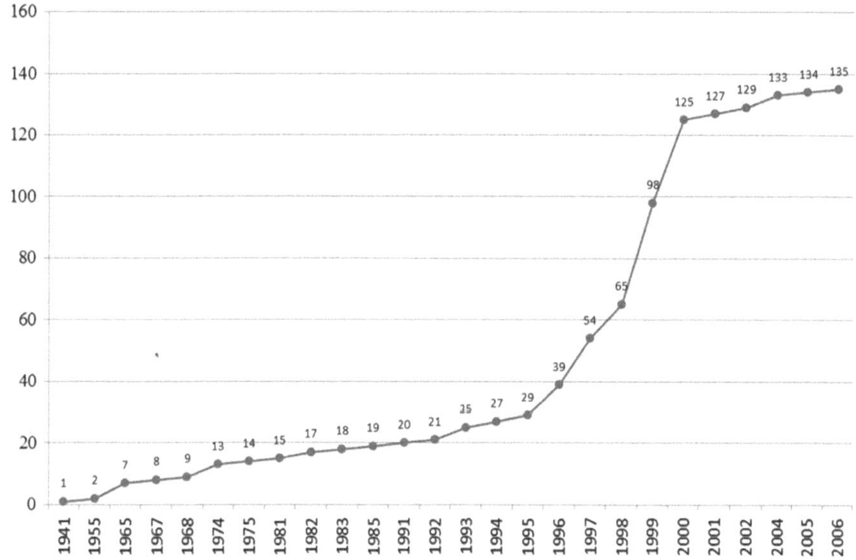

Fig. 1 Number of Teacher Preparation Institutes (*IFD*) in years when new ones were created. *Source* MEC—Viceministerio de Educación Superior, Abril 2012

This phenomenon merits special attention, above all to locate the main problems faced in initial teacher preparation for the elementary level. Figure 1 shows the growth of the IFD. In the mid' 1990s, as the period in which democracy was reinitiated, the growth spurt in the number IFD was also initiated. By 2006, the Ministry of Education had registered 135 IFD: 41 were public and 94 were private (MEC, 2012).

Faced with the excessive increase in private *IFD*, the Ministry of Education and Culture suspended the further creation of both public and private *IFD* via Resolution N° 10747/00. Later, in 2007 (MEC, 2000), the preparation of elementary teachers was also suspended. With this later suspension, there was an abrupt closing of many *IFD*, particularly in the private sector.

The suspension of the opening of new *IFD* and of the preparation of elementary teachers, was occasioned by the oversupply and resulting unemployment of many teachers. At the same time, there was a significant undersupply of lower secondary teachers (Secretaría Técnica de Planificación, 2014).

Available data indicate that in 2011 there were 41 public *IFD* and 80 private that were preparing lower and upper secondary teachers. The *IFD* were also providing in-service educator professional development (Rojas, 2011).

Structure of the Paraguayan Education System

Beginning in the 1990s, the State gradually increased investments in public educational institutions. This was necessary, in order to meet the need to offer free and compulsory basic public education that was increased from six to nine years.

In 2010, compulsory and free education was increased to include the preprimary and upper secondary levels, in order to help achieve the goals of the National Plan for Education 2024. Among the promises of that plan were to guarantee equality of opportunities through access; and to improve quality, efficiency and equity in Paraguayan education as a public good and fundamental right.

Today, Paraguayan education is composed of three levels. Level 1 includes Preprimary (*Educación Inicial*) and Elementary and Lower Secondary (*Educación Escolar Básica*). Level 2 is Upper Secondary (*Educación Media*). Higher Education is called Level 3.

The 1998 General Law of Education made Elementary and Lower Secondary, as well as Preprimary, compulsory. The process of including Preprimary as free and compulsory is still in progress. The Ministry of Education and Culture's web portal (MEC, 2017) is not particularly clear with respect to the different modalities, stages and cycles of Preprimary, indicating that "it will systematically belong" to Basic School Education (*EEB*) along with Elementary and Lower Secondary.

Table 1 shows the structure of the first two levels of education in Paraguay, along with sublevels, grades and corresponding ages of the students.

The Ministry of Education and Culture determines how many hours of Mathematics instruction is required in all educational institutions in the country. As can be seen below in Table 2, there are more required hours in the early grades.

For Upper Secondary, the Ministry of Education and Culture has determined distinct numbers of hours of Mathematics depending on the particular programs of study, as is shown in Table 3.

The structure of Level 3 (higher education) includes both universities and non-university tertiary institutions. Table 4 indicates the higher education sublevels, authorized institutions, programs offered and program characteristics with special attention to teacher preparation programs (CNP, 2013, MEC, 2009).

Initial Teacher Preparation

Details on the current status of teacher preparation, programs of initial preparation and specific programs for preparing Mathematics teachers are presented below.

Table 1 Structure of the Paraguayan education system: Levels 1 and 2

	Level	Sublevel	Grades	Ages
Level 1	Preprimary (informal)	Not in school	Daycare (*Casa del Niño)*	3 to 5 years old
	Preprimary (Formal)	Cycle 1	Daycare (*Jardín Maternal*)	Birth to 2 years old
		Cycle 2	Pre-school (*Jardín de Infantes*)	3 to 5 years old
			Kindergarten	5 years old
	Basic school education (*EEB*)	Cycle 1 of *EEB* (Primary)	Grades 1 to 3	6 to 8 years old
		Cycle 2 of *EEB* (Upper elementary)	Grades 4 to 6	9 to 11 years old
		Cycle 3 of EEB (Lower secondary)	Grades 7 to 9	12 to 14 years old
Level 2	Upper secondary	Scientific with an emphasis in: Letters & Arts, Social Sciences, or Basic Sciences & Technology	Courses 1 to 3 (Grades 10 to 12)	15 to 17 years old
		Technical: Administration, Informatics, Industrial, Services, etc.	Courses 1 to 3 (Grades 10 to 12)	15 to 17 years old

Source The authors with data from the Ministry of Education and Culture

Table 2 Hours of mathematics per week—elementary and lower secondary

Educational level	Cycle 1 of *EEB* Primary (Grades 1, 2 and 3)	Cycle 2 of *EEB* Upper Elementary (Grades 4, 5 and 6)	Cycle 3 of *EEB* Lower Secondary (Grades 7, 8 and 9)
Hours per week	8	5	5

Source The authors with data from the Ministry of Education and Culture

Current Status of Teacher Preparation in General

Law N° 1.725,—01 (Poder Ejecutive del Paraguay, 2003), that established the Educator Statute, in Article 2 states that: "In Paraguay, a professional educator is a person that possesses a certification in any humanistic, scientific or technological discipline who is regularly dedicated to teaching activities in educational establishments, centers or institutions, and has been duly registered." Nevertheless, as is indicated in

Table 3 Hours of mathematics per week—upper secondary

Sublevel			Course 1 (Grade 10)	Course 2 (Grade 11)	Course 3 (Grade 12)
Scientific	Program				
	Common required courses		5	4	3
	Additional according to emphases				
		Basic Sciences	4	3	
		Social Sciences	2		
Technical	Program				
	Common required courses by specialties		3–5	3–5	2–4

Source The authors with data from the Ministry of Education and Culture

Table 4 Structure of the Paraguayan education system: Levels 3 (higher education)

Sublevel	Authorized institutions	Programs	Characteristics
Non-university tertiary level	Tertiary non-university institutes	Teacher preparation	Teachers for Cycles 1&2 (primary & upper elementary). Duration: 3 years
			Teachers for Cycle 3 (lower secondary), with areas of specialization. Duration: 3 years
			Teachers for upper secondary, with areas of specialization. Duration: 3 years
		Higher technical education	Various non-university tertiary technical degrees (*tecnicaturas*). Duration: 2 or more years
University tertiary level	Universities and Higher Institutes	Undergraduate	Higher professional preparation. Bachelors degrees (*licentiaturas, ingenierías,* etc.)
		Graduate	Specializations, masters, doctorates

Source The authors with data from the Ministry of Education and Culture

Table 5 Percentage of certified teachers by education level

Level	Certified (%)	Uncertified (%)
Preprimary	49	51
Elementary	92	8
Lower secondary	63	37
Upper secondary	53.0	47.0

Source MEC, (2012)

Table 5, many teachers at most educational levels do not meet the conditions established in Article 2.

Table 5 indicates the percentages of certified teachers. It can be observed that the greatest number of certified teachers are working at the elementary school level followed by teachers working in lower secondary. It should be noted that those are the levels that have been obligatory within the Paraguayan education system for the longest time.

At the upper secondary level, of the 47% that are not certified 40% are certified at other levels and 7% do not have a higher education degree (Banco Mundial, 2008 cited in Banco Mundial, 2013, p. 43). The curricular design for Upper Secondary Education (2002–2004) proposes an organization of the disciplines within general areas. However, in the administrative organization teachers are hired by discipline. Teacher preparation within general areas just began in 2002, at the same time as the implementation of the reforms at that level. To date, the Teacher Preparation Institutes have produced only 10 cohorts of teachers prepared to implement curriculum by general areas (Banco Mundial, 2013, p. 43).

On the other hand, in rural areas there are many teachers without an adequate preparation and, in urban areas there are more teachers with the necessary preparation in private schools than in public schools (Falus & Goldberg, 2011).

In Table 6 it can be observed that for lower secondary schools it is in the cases of Guaraní and Mathematics that the greatest percentages of teachers are not certified for the courses they are teaching.

Program for Initial Teacher Preparation

Today, universities, higher institutes and post-secondary professional preparation institutes (specifically teacher preparation institutes) can prepare teachers. Table 7 indicates the number of such institutions both public and private.

According to the Ministry of Education and Culture (MEC), and Together for Education (*Juntos por la Educación*), (2013):

> The Teacher Preparation Institutes (*Institutos de Formación Docente-IFD*) are Post-Secondary Professional Preparation Institutes, a component of the Higher Education subsystem, that prepare teachers for Preprimary, Basic (Primary, Upper Elementary, Lower Secondary) and Upper Secondary, as well as, in-service education. They are not autonomous.

Table 6 Number of teachers certified by general areas for lower secondary schools

	Certified	Uncertified
Communications Area		
Spanish Language and Literacy	3300	2800
Guaraní Language and Literacy	1900	4800
Art Education	3600	1900
Social Scientific Area		
Mathematics	2300	3000
History and Geography	3400	2600
Ethics and Civics	3700	2600
Natural and Health Sciences	4400	3100
Work and Technology	4500	1400
Physical Education	2900	1900
Community Development Area		
Community Project	600	9000

Source MEC (2009)

Table 7 Number of public and private teacher preparation institutions in Paraguay

	Public	Private
Teacher Preparation Institutes	40	94
Higher Institutes	8	30
Universities	8	44
Total	56	168

Source Mesa Temática de Formación Docente, MEC, Juntos por la Educación, (2013)

They depend upon the MEC for approval to fill positions and for program approval. As of 2012, the MEC had registered 40 public and 94 private *IFD*. Regional Education Centers (*Centros Regionales de Educación-CRE*), that were originally created to encourage experimentation and innovation, are institutions that serve a large student population beginning with preprimary, and also have teacher development programs related to school activities. There is a total of seven *CRE*. They are all publicly funded as Teacher Preparation Institutes (*IFD*).

The Law of General Education 1264/1998 passed by the National Congress, states that the *IFD* should (a) provide teacher development with the highest possible professional, scientific and ethical quality; (b) achieve an effective performance of the profession in all levels of the education system and in the diverse modalities of education activity; (c) provide in-service teachers with professional development; and, (d) strengthen the competency in the field of research and in the development of theory and practice of Education Sciences.

The Teacher Preparation Institutes (*IFD*) have historically implemented two admissions mechanisms:

- Up to 2005, an admissions test for which students prepared by taking month-long daily classes in Mathematics, Spanish and Guaraní. The test was designed by the General Directorate of Teacher Preparation of the Ministry of Education and Culture for admission to all public and private *IFD* in the country;
- Beginning in 2006, there has been a new test preparation course for admission to teacher preparation programs. That preparation course lasts 420 h with four modules of 105 h each. To pass the course students must score a minimum of 60% for each module.

There are currently four different programs (called *Profesorados*) for initial teacher preparation: Preprimary, Elementary, Lower Secondary (with a major in various disciplines, including Mathematics), and Upper Secondary (also with a major in Mathematics). All programs are for three years, except Lower Secondary that can be completed in one year if a program in elementary education has already been completed.

There are also teacher preparation programs (usually developed by the *IFD*) for in-service teachers who have just a high school diploma. There are various programs for those teachers, offered as hybrid courses, that lead to the following degrees:

- Certified Elementary Teachers.
- Certified Lower Secondary Teacher.
- Certified Upper Secondary Teacher with a major in a specific discipline (Spanish, Guaraní, Ethics and Civics, History and Geography, Mathematics, Natural and Health Sciences).

For admissions to public or private universities and higher institutes it is necessary to be a high school graduate and pass an admissions test. Students who successfully complete all graduation requirements receive certification as a teacher as well as a Bachelors (*Licenciatura*) degree in the relevant discipline or an Education Sciences degree with a major in the discipline.

Preparation of Teachers for Preprimary Education

Programs for the preparation of preprimary teachers at the *IFD* and universities have been undergoing curricular changes. The need for the changes is largely a result of the ministerial decision to make preprimary education mandatory, and a desire to improve both quantity and quality of teachers for that level.

In 2006, Ministerial Resolution N° 114.415/06 implemented the curriculum for Specialization and Professionalization in Preprimary Education at 27 *IFD*. In 2009 that curriculum was revised, evaluated and modified (Rojas, 2011).

A reality with respect to students who wish to enter teacher preparation programs is their relatively weak academic backgrounds. The main cause of the weakness is the quality of education they have received which is linked directly to their socio-economic backgrounds. Students that choose teacher preparation programs tend not to be the best students and select education because they cannot gain admission to other university programs. Also, they need quick access to a job to satisfy their economic needs. To confront this situation, some teacher preparation institutions are basing admissions decisions on attempts to determine vocational and attitudinal aspects related to working with children. Academic remedial programs have also been put in place to try to alleviate student academic deficiencies. However, it is not known what the results of such efforts have been because no significant evaluation process has been carried out.

Preparation of Teacher for Elementary Education

In Paraguay, there are two types of institutions charged with preparing Elementary teachers:

- The Teacher Preparation Institutes (*IFD*), both public and private, that are regulated by the Ministry of Education and Culture. Most future elementary teachers attend public *IFD* in urban areas.
- The Higher Institute of Education that is academically autonomous and thus able to develop its own program and courses.

In the *IFD,* the curriculum for future elementary teachers still does not include the development of competency-based education. Since both lower and upper secondary teacher preparation programs do include competency-based education, there is a disconnection with what is being implemented in elementary as compared to secondary classrooms (Rojas, 2011).

Teacher Preparation for Secondary Education

Currently there are four modes or pathways to certification in secondary education:

- Certification in Secondary Education with a Bachelors (*Licenciatura*) degree in Education Sciences, offered by universities and higher institutes of education,
- Alternative certification offered by universities, higher institutes of education, teacher preparation institutes (*IFD*) and regional centers (a system of sequential preparation: specialization followed by professional preparation),
- Certification in Secondary Education, a concurrent program offered by the higher institutes of education, the teacher preparation institutes (*IFD*) and regional center, and
- A bachelors (*licenciatura*) in Education Sciences with the various majors.

The curricular design for Upper Secondary Education (2002–2004) proposes an organization of the disciplines within general areas. However, as was mentioned previously, in the administrative organization teachers are hired by discipline. The preparation of teachers by general area was initiated in 2002, at the same time as the implementation of curriculum by general areas at that level. By 2009, six cohorts of teachers prepared by general areas at the teacher preparation institutes (*IFD*) had graduated (MEC, sept, 2010 cited in Banco Mundial, 2013, p. 53). In the universities, upper secondary teachers are usually prepared for specific disciplines.

Very few of the teachers prepared to implement curriculum by general areas are working in the formal education system.[1] Also, the General National Budget does not include categories based on general areas, so the contracting by teaching hours according to specific disciplines is still in effect (MEC, sept, 2010 cited in Banco Mundial, 2013, p. 53).

Initial Preparation of Mathematics Teachers

As has been described previously, preprimary and elementary teachers are prepared to be generalists with no emphasis or focus of attention on Mathematics Education. Recently, there have been programs that offer a major in Mathematics for lower and upper secondary teachers.

Numerous universities in the country offer an undergraduate degree in Education Sciences with a major in Mathematics. Given that the process of program accreditation is just beginning in the country, it is difficult to have exact and reliable information about such programs.

Up until now, there has not been an effective system of accreditation. The various institutions that offer programs in Mathematics Education have existed without interinstitutional tensions. That may change as the accreditation processes are implemented. Differences in the quality of the various programs may become evident as that quality is used as an indication of the ability of graduates to be successful in classrooms.

However, it can be said that the mathematical content in the Education Sciences degree with a major in Mathematics does not reach a minimum level. It is never more than 50% of the total curriculum and it is often only two modules of the total.

Among the institutions that offer majors in Mathematics Education, are the Higher Institute of Education "Dr. Raúl Peña"; the Faculty of Exact and Natural Sciences (*FACEN*) at the National University of Asunción and the Faculty of Exact Sciences and Technologies at the National University of Concepción.

In general terms, the programs offered by each of the three institutions have the following academic components: (a) general education and other disciplines (Letters, Natural Sciences, Social Sciences, Languages, Technology, Art, elective courses, etc.); (b) Mathematics; (c) general pedagogy (pedagogical theories, general teaching

[1] A public competitive system for applying for teaching positions has been in operation since 2004.

Table 8 Distribution of credits and hours in the Bachelors in Education Sciences at the *ISE*

Field of Study	Hours	Percentage (%)
Pedagogy	768	24
General Education	512	16
Student Teaching and extension	384	12
Specific content courses	1536	48
Total classroom hours	3200	100
Total non-classroom hours	350	
Total theoretical-practical hours	3550	

Source The authors

methods, Philosophy of Education, Educational Planning, etc.); (d) Mathematics Pedagogy; (e) Student Teaching; and (f) Research in Mathematics Education.

Below are descriptions of the offerings at each of those three institutions.

Higher Institute of Education "Dr. Raúl Peña". The Higher Institute of Education (*ISE*)[2] "Dr. Raúl Peña", is administratively controlled by the Ministry of Education and Culture, but has pedagogical autonomy. Its programs for initial teacher preparation are designed in collaboration with specialists from the Ministry. The official curricular contents for teacher certification programs should be developed in their entirety by the public and private Teacher Preparation Institutes (*IFD*), Higher Institutes of Education (*ISE*) and universities. They can also incorporate other contents beyond those officially required.

Bachelors (Licenciatura) in Mathematics Education (ISE—Dr. Raúl Peña). As soon as Law 1692, 2001 (CNP, 2001) recognized this institution as a Higher Institute of Education (*ISE*), programs were created in specific areas with adjustments to curriculum and the total number of hours. Thus, the programs were elevated to the level of an undergraduate degree based on criteria that seek relevance and consistency in professional preparation. The curricular guidelines and adjustments were developed exclusively by the faculty at the *ISE* without consultation with external sources. Members of that faculty included individuals with undergraduate degrees in Mathematics and graduates of the *ISE* itself.

One of the new programs was the Bachelors in Mathematics Education which targeted high school graduates with an interest in becoming Mathematics teachers. Graduates of the program were certified as Mathematics teachers (Table 8).

The characteristics of that program are indicated below:

- Length: 4 years (3550 academic hours)
- Organization: 8 semesters
- Types of courses: Theoretical-Practical (Table 8)

[2]The data on the *ISE* were obtained from faculty members and its website http://www.ise.edu.py.

Table 9 Distribution of credits and hours in Bachelors in Mathematics Education program at *FACEN*

Academic component	Total credits [a]	Percentage of credits (%)	Total hours
General Education and other courses	20	15	459
Mathematics	76	57	1700
General Pedagogy	22	16	544
Methods of Teaching Mathematics	3	2	68
Student Teaching	8	6	204
Research in Mathematics Education [b]	5	4	136
Totals	134	100	3111

Source The authors
[a]1 credit is equivalent to 1 class hour of theory per week or 2–3 h of lab per week
[b]The program has two course with this name. In the first, theoretical aspects of scientific research are developed. In the second, a research project is carried out and defended

- Fields of Study: Pedagogy, General Education, Student Teaching, Specific Content Courses
- Number of Courses: 57
- Classroom Hours: 25 per week
- Non-Classroom Hours: 200 h of Student Teaching and 150 of Extension.

Most of the courses in General Education and Specific Content courses are taught by professors with an undergraduate degree in Mathematics.

Faculty of Exact and Natural Sciences (*FACEN*) at the National University of Asunción. The following programs at FACEN are related to Mathematics Education:

- Bachelors (*Licenciatura*) in Mathematics Education (face-to-face): created in 2009 to prepare teachers with the mathematical knowledge and pedagogical abilities to teach at the lower and upper secondar levels.
- Bachelors (*Licenciatura*) in Mathematics Education (hybrid): begun in 2011, it has the same curriculum as the face-to-face version.

To recruit students in its undergraduate program in Mathematics Education, (both the face-to-face and hybrid) *FACEN* uses it webpage and social networks. It also does direct recruiting in some upper secondary schools. For admissions students must have completed upper secondary and meet admissions requirements set by the Faculty.

The Bachelors in Mathematics Education is a four-year degree. It leads to certification to teacher at the lower and upper secondary levels.

The system of credits that facilitates students to organize individual Programs of Study can be seen in Table 9.

The courses in General Education are Pedagogy (I and II), Philosophy of Education, Educational Psychology, Sociology of Education, Education and New Technologies (I and II), Curriculum, Spanish, Guaraní, Assessment of Learning, Technical English, Methods of Teaching Mathematics, Research in Mathematics Education (I, II and III), Student Teaching in Mathematics (I, II and III), Professional Ethics, and Educational Administration.

The specific content courses are Geometry and Trigonometry, Elementary Algebra, Arithmetic, Analytic Geometry, Mathematical Logic, Algebra, Calculus of One Variable, Set Theory, Calculus of Various Variables, Linear Algebra, Probability and Statistics (I, II and III), Educational Planning, Differential Equations, Numerical Methods, Financial Mathematics, Demography, Project Control, Abstract Algebra (I and II), and Linear Programming (UNA, 2012).[3]

Both the General Education courses and the specific content courses are usually taught by professors with an undergraduate degree in Mathematics from *FACEN*. They also have a specialization in University Teaching which is a fundamental and mandatory requisite for teaching at the university level.

Faculty of Exact Sciences and Technologies at the National University of Concepción (*UNC*). The programs related to Mathematics Education in the Faculty of Exact Sciences and Technologies[4] are the following:

- Bachelors (*Licenciatura*) in Mathematics and Physics: a four-year program that began in 2012, and
- Bachelors (*Licenciatura*) in Applied Mathematics: a 2014 restructuring of the four-year Mathematics and Physics degree with the five-year Civil Engineering degree.

For admissions to both programs students must pass a year-long Entrance Preparatory Course (*CPI*) with courses in Arithmetic, Algebra, Geometry and Trigonometry, Analytic Geometry, Differential and Integral Calculus, Physics, Communication and Learning Methodology. Both programs also provide elective courses that enrich the curriculum with pedagogical content needed to be certified as a teacher (Tables 10 and 11).

The Continuing Preparation of Teachers

According to the Ministry of Education and Culture (MEC, 2013), in Paraguay, Continuing In-Service Teacher Preparation should integrate, complete and deepen Initial Teacher Preparation. As such it should permit, update, enrich, and "deepen knowledge, generate innovations, promote processes of transformation and give answers to the requirements of self-development. It is oriented to the improvement of edu-

[3]The National University of Asunción, Higher University Council, Resolution No 0144-00-2012.

[4]The webpage of the Faculty of Exact Sciences and Technologies is http://www.facet.unc.edu.py/.

Table 10 Distribution of hours in the degree in Mathematics and Physics at *UNC*

Field of Study	Hours	Percentage (%)
Specific scientific courses	2784	77
Fundamental humanistic-pedagogical courses	720	20
Complementary courses	96	3
Total theoretical-practical hours	3600	

Source The authors

Table 11 Distribution of hours in the degree in Applied Mathematics at *UNC*

Field of Study	Hours	Percentage (%)
Basic	1140	39
Professional	1380	47
Complementary	240	8
Elective	180	6
Total theoretical-practical hours	2940	

Source The authors

cation and the deepening of levels of professionalism. It offers the possibility of development, updating, professionalization and specialization" (MEC, 2013).

The Ministry determines the characteristics that identify each one of these possibilities as follows:

- Development: courses of at least 100 h that permit the strengthening of specific aspects, such as assessment, teaching methods or research,
- Updating: courses that look to incorporate innovative pedagogical competencies,
- Professionalization: courses directed at teachers that are not certified in certain areas or levels of professional endeavor, and
- Specialization: courses for those who already have a university degree and desire to specialize in a certain area.

The Legal Framework and Structure of Continuing Preparation

Executive Order 468/2003 regulates the application of Law Nº 1725/2001 as it relates to the teaching profession at the preprimary, elementary and secondary levels of the National Education System. Chapter 2, Section A, Article 5 of the Executive Order determines that continuing teacher preparation courses include courses for pedagogical preparation for university graduates without teaching degrees; courses for certification for teachers who are high school graduates without higher education degrees; and specialization courses and other courses, seminars and workshops that are oriented to improving teaching.

Article 6 of Executive Order Nº 468/2003 also establishes that it is the Ministry of Education and Culture, in coordination with departmental and municipal governments, and other interested organizations, that promotes continuing preparation.

Likewise, Article 7 establishes that the Ministry of Education and Culture is responsible for offering programs of continuing preparation, without cost and for all teachers. It also indicates that for career advancement teacher must pass a minimum of 100 h of continuing preparation courses.

Offering Continuing Preparation

Historically in Paraguay, teacher continuing preparation in the form of seminars, workshops and meetings has been sporadic and short. Such events were usually offered by the Directorate of Continuing Teacher Preparation of the Ministry of Education and Culture. Teacher participation was usually voluntary. Although it was free, the lack of time and transportation expenses were causes for teachers not to participate.

Courses and workshops for continuing preparation have usually been short. When they have fulfilled certain prescribed requirements, like for example a minimum of 100 h, a certificate might be awarded and it could serve for career advancement.

According to data from the Ministry, 1710 teachers participated in Teacher Specialization programs in 2009. It is important to point out that Teacher Specialization programs are for continuing preparation and consist of short courses for development and specialization. They are directed at teachers in schools in distinct parts of the country.

These specialization courses have not had a direct impact on the career ladder. In fact, every five years points are assigned and it is possible for teachers to be placed on the career ladder upon presenting the relevant documentation. During many years, and despite the career ladder resolution, teachers did not receive salary increases because the Ministry did not have the necessary resources. However, during 2010–2011 there were resources, and teachers who accepted advancement on the career ladder did receive salary increases.

In the period from 2000 to 2012, the Inter-American Development Bank financed a program called the Hekokatuva Living School (*Programa Escuela Viva Hekokatuva*). It implemented a series of strategies promoted by the National System of Teacher Development (*SINAD*). One of the strategies was "learning circles". The strategies were defined as mechanisms for permanent development in which interchange and learning in pairs was promoted.

Among the offerings of continuing preparation that should be highlighted, particularly for their regularity, are those of the Faculty of Exact and Natural Sciences (*FACEN*) at the National University of Asunción. Every year free continuing preparation 100 h courses are offered in two modalities: face-to-face and hybrid. For elementary teachers, there are courses on Arithmetic, Geometry and Statistics. For lower secondary, there are Algebra and Geometry courses. For upper secondary, a variety that includes Algebra, Trigonometry, Analytic Geometry and Calculus has been available. There are also continuous preparation courses in Basic Sciences and technology. This offering of a wide variety of courses began in 2010 and as of 2015 over 4000 teachers from all over the country had participated.

In concluding, it can be mentioned that the continuing preparation offerings, from the Ministry and other institutions, respond to the criteria of the institution making the offering. Each institution defines both the content and the approach that will be taken. Although it is true that the courses often reflect principles and foundations established by the Ministry, the different offerings are totally unrelated to each other.

The Evaluation of Continuing Preparation

There is only limited information on the impact that continuing teacher preparation has had on classroom practice. Findings from a research study carried out by the Higher Institute of Education "Dr. Raúl Peña" (ISE, 2008 cited in Banco Mundial, 2013, p. 65) on the performance of their graduates show that only 3 of 10 graduates pursued continuing preparation. Various factors seem to explain that situation, particularly personal and financial reasons. Excessive work schedules, given that most teachers work in more than one school, and low salaries, impede the investment in continuing preparation are the arguments that were most often heard.

The conclusions to that study present evidence that the majority of the graduates that do participate in continuing preparation take courses related to Educational Assessment. That result seems to be consistent with another finding related to the reported difficulties in implementing pedagogical strategies and assessing students (ISE, 2008 cited in Banco Mundial, 2013, p. 65).

As a conclusion, it can be said with respect to continuing teacher preparation in Paraguay that, despite being a necessity, it is being presented by the Ministry to patch weaknesses and not as part of a continuum from initial to continuous preparation. Martínez (2013) affirms this conclusion.

Recent Notable Actions in Initial and Continuing Preparation

Previous studies have shown that the proposals developed since 2004, with the design of new curricula for initial and continuing preparation, have focused on new ways of teaching and of conceiving education. Having those who would be providing teacher professional development experience the same demands and experiences as would be expected of teachers, led to an innovative concept of teacher development. This advancement was noted in an evaluation carried out by the Inter-American Development Bank in 2010 as it reflected changes in classroom practice (Martínez, 2013).

Martínez (2013) also documented the process of teacher professional development that took place from 2004 to 2008 after the creation in 2004 at the Ministry of Education and Culture of the Directorate General of Human Resources. The purpose of this directorate was to make hiring decisions for education. Competency tests were given to candidates for teaching and administrative positions in publicly-financed schools. Since 2009 a National Campaign of Pedagogical Support for Teachers and Administrators in Educational Institutions has been implemented. Its purpose has been to respond to learning needs by levels and prioritized needs.

In 2011 as part of this campaign in 2011 there was a professional development program for upper secondary Mathematics in which 1200 teachers of the national total of 3000 participated. It took place in 47 cities simultaneously. There were five subjects: Trigonometry; Matrices and Determinants; Analytic Geometry; Functions; and Limits, Continuity, Derivatives and their Applications. During the presentation of these courses there were attempts to include not only the mathematical content, but a diversity of strategies for teaching the content (MEC, 2011). This approach, proposed and executed by OMAPA, was unprecedented and greatly appreciated by the teachers.

Evaluations of the mathematical knowledge participating teachers showed a 10% increase from pretest to posttest. The success of this initiative shows the interest that teachers have in receiving courses in areas where they know they need strengthening. However, this campaign was impacted by changes in the Government and was discontinued.

Research in Mathematics Education and Its Relation to Teacher Preparation

One of the critical aspects of teacher preparation in Paraguay is the lack of capacity to produce relevant research. There is also little capacity for utilizing the little research that has been accumulated to improve practice.

Although at present, research is almost nonexistent, the prospects for the future are auspicious.

The undergraduate degrees in Mathematics Education at the *ISE* and at *FACEN* include courses on research in their curricula. For example, in order to graduate, students must complete a research thesis. However, there are no criteria or guidelines for thesis advisors. Therefore, as in the past, students are generally on their own in the selection of the focus of their research. Since the programs are new, it can be hoped that results will be seen that can facilitate research.

Two recently implemented Masters degree programs could be important in the near futures and are detailed below.

The National University of Asunción through *FACEN* has been offering a Masters in Mathematics since 2006. Its goal is to contribute to the development of research in pure and applied Mathematics, and to the qualitative improvement of science and technology. It is the first Masters in Mathematics in Paraguay. Although it is in Mathematics, not in Mathematics Education, it is necessary to mention it in this report because it offers the highest level of preparation in Mathematics in the country in all of its history.

Since September of 2015 the National University of Concepción has been offering a Masters in Science Teaching which aims to prepare teacher-researchers with competencies in developing research in Mathematics Teaching, Physics Teaching and Chemistry Teaching. Thus, these new researchers will contribute to the identification and generation of proposals for solving educational problems in the country. A contextualized and relevant theoretical body of knowledge on the teaching of Mathematics, Physics and Chemistry is being built. The research areas established by this graduate program include Teacher Preparation; Problem Solving in Teaching and Learning; Curricular Design, Development and Evaluation; Technological Learning Environments; Teaching of Specific Disciplines; all in the context of Basic Sciences. It is the first Master's degree program that will be producing research in Mathematics Education.

It is noteworthy that the Faculty of Humanities and Education Sciences at the National University of Concepción from 2012 to 2015 was encouraging its students to present projects on Mathematics Education. Among the titles were "Strategies for teaching Mathematics," "Strengths and weaknesses of primary education," "Factors that influence learning in upper elementary," "Implementation of a Logic and Mathematics Corner in elementary," "Implementation of a heuristic methodology in Mathematics problem solving in lower secondary," "Teacher evaluation practices based on the learning process and technology in upper secondary Mathematics classroom," and "Factors related to the low Mathematics achievement of students in upper secondary technical schools."

Challenges and Opportunities for Teacher Preparation in Paraguay

As has already been pointed out, and like other countries in Latin America, teacher preparation programs in Paraguay have had little impact on classroom practice. The emphasis on teacher preparation continues to be theoretical without sufficient specific pedagogical practice. This situation in Paraguay had already been pointed out twenty years ago by Reimers (1993). Moving the Teacher Preparation Institutes to the higher education level in 1974 has not led to relevant positive results.

Martínez (2013) presented interesting reflections with respect to the critical junctures faced by both initial and continuing teacher preparation. Findings from programs developed with Inter-American Bank funding, and evaluations carried out in

2004 and 2010 also registered the same critical junctures that have not been overcome. A study published by the Regional Office for Education in Latin America and the Caribbean of UNESCO (2012) indicates that the situation in Paraguay coincides with other Latin American and Caribbean countries.

Students entering the Teacher Preparation Institutes (*IFD*) have enormous difficulties related to abilities not developed in secondary school. They are not prepared to meet the demands of teacher preparation in priority areas such as Spanish, Guaraní, Mathematics and general culture. The basic abilities of students who enter teacher preparation programs are related to the poor quality of the Paraguayan education system which is rated as one of the ten worst in the world. Such weaknesses impose restrictions when preparing teachers. Students have difficulty with academic achievement for at least two main reasons: their socioeconomic origins and the quality of education they have received. At the same, those entering teacher preparation programs are not the best students. Rather, they are those who have not been admitted to a university and need to get a job to earn a living. That is, entering the teaching profession is a response to economic necessity and the search for a quick route to a job. The best students select other professions that have more social prestige and better salary prospects, such as engineering, medicine, law and economics.

Many important weaknesses are detected among students who enter initial teacher preparation programs. They lack the basic abilities that are needed for success. They have difficulties with their relationships with their professors and do not possess the minimum conditions of the *IFD* needed to fulfill their roles as students.

Continuing preparation is not treated by the Ministry of Education and Culture as a relevant aspect of teacher preparation policy. Instead, it is seen as a tool to fix weaknesses, and no relation is established between initial and continuing preparation. Also, there is limited evidence about the impact that continuing preparation has on classroom practice.

The limitations in both initial and continuing teacher preparation in Paraguay are related to the socioeconomic situation of pre-service and in-service teachers. The majority come from middle and lower classes. Teachers dropout of the profession for multiple personal, social, economic and health factors.

The Ministry has put effort and care into the development of curricular design for initial and continuing preparation for the preprimary, lower secondary and upper secondary levels. The preprimary programs have been evaluated and modified. The lower secondary programs have not been implemented. Nevertheless, the lower secondary programs are related to and coherent with the preprimary programs. Upper secondary programs need to be modified after an evaluation. The curriculum for future elementary teachers has been modified. Nevertheless, it will be necessary to wait until a cohort of students has graduated to be able to evaluate how well they perform.

The academic preparation of the teacher education professors is one of the most important factors in ensuring the quality of teacher preparation. It should be noted that the teacher education professors usually do not have graduate degrees. The most negatively affected areas are Mathematics and the Natural and Social Sciences.

Table 12 Results from written tests used in the teacher and administrator selection process

Year	Month	Number Evaluated	Number Passing	Percent passing (%)
2009	February	7000	910	13
2009	May	9500	2185	23
2010	February	4998	1399	28
2010	May	9034	4554	50
2011	February	8085	2742	34
2012	February	8118	3301	41
2013	February	6862	1816	27

Source MEC, Juntos por la Educación. Mesa Temática. Formación Docente 2012–2013

One of the reasons for the failure of many to pass the tests given as part of the Ministry's application process for teaching positions, is inadequate teacher preparation. There are many teacher preparation programs, but they have limited coverage, are discontinuous and lack a systemic approach (Rivarola, 2006).

The results from the tests given to those applying for teaching and administrative positions, since 2009, show deficiencies in the preparation of teachers and administrators. Most applicants failed the tests and were not eligible to be hired. The results of the test to the applicants are presented in Table 12.

The data in Table 12 support the need to concentrate attention on the initial and continuous preparation of teachers, in order to foresee better learning, in the medium and long term, for the children and youths of Paraguay.

Reports from the Ministry of Education and Culture in 2010 pointed out that in many Paraguayan schools the teachers do not have preparation specific to the level in which they are teaching. At the elementary level 55% of the teachers are not duly certified. The uncertified teacher percentage for lower secondary is 38% and it is 47% for upper secondary. (MEC, Juntos por la Educación, 2013).

One new opportunity is the decision by the Higher Institute of Education to elevate the preparation of teachers to the level of Bachelor's degree in 2012. The first cohorts will soon be graduating.

Another recent and promising development with respect to initial teacher preparation is that the Teacher Preparation Institutes (*IFD*) are being evaluated. Leaks from meeting of Ministry experts and others suggest that only six of the 40 *IFD* meet the minimum required academic conditions. The official diagnostic report and recommendations with respect to this situation will initiate radical and profound changes that are needed.

Teachers are calling for future professional development that begins by filling gaps in content knowledge that they missed in their initial preparation. Next, they want a specific focus on teaching Mathematics. Finally, they would like more help incorporating technology into their teaching and would appreciate an introduction to some areas that are still new to Mathematics Education in Paraguay, such as problem solving and modeling.

With respect to research, the most relevant anticipated changes should come from three new programs: *PRONII, PROCIENCIA* and *BECAL*.

The National Program for Incentivizing Research (*PRONII*) has been implemented by the National Council for Science and Technology (*CONACYT*) since 2011. Its goal is to strengthen and expand the scientific community in the country. It seeks to promote the careers of researchers by registering them, evaluating their scientific and technological production, and providing them with economic incentives. Up until 2013, development had barely begun. Beginning in 2014 with the creation of the Fund for Excellence in Education and Research (*FEEI*), *CONACYT* began receiving funds to strengthen *PRONII*, and to create and execute the Program for the Development of Science and Technology (*PROCIENCIA*). Also, with these funds it has been possible to create the National Program for Graduate Study Abroad to Strengthen Research, "Don Carlos Antonio López" known as *BECAL*.

The importance of these funds and the results that they are already producing, merit a special section.

Fund for Excellence in Education and Research (FEEI)

The extraordinary initiative called "Paraguay Now" organized by citizens with committed journalists deserves special mention. Their efforts have led to the creation of a fund dedicated to providing resources to elevate the quality of education in the quest for educational excellence and the promotion of research and development.

The funds are made available through Law N° 4.758/2012 (CNP, 2012) which directs to the National Fund for Public Investment and Development (*FONACIDE*) a significant portion of payment from Brazil for hydroelectric energy generated by Itaipú Dam. The Fund for Excellence in Education and Research (*FEEI*) receives 30% of the generated resources (FEE, 2014).

Article 12 of Law N° 4.758/2012 established the priorities for the investment of this fund. Those priorities are summarized below:

a. Programs for incorporating ICT (Information and Communication Technologies);
b. Support to programs for improving initial teacher preparation;
c. Programs for improving educational opportunities in public schools and teacher preparation institutes, through the implementation of infrastructure and equipment projects,
d. Programs for comprehensive early childhood care;
e. Programs to organize, form and strengthen cooperation networks for parents and students´ tutors;
f. The awarding of scholarships for higher education study within the country and abroad;

g. Research and development, that consists of programs and projects proposed by the National Council for Science and Technology (*CONACYT*) mainly for the Program of Incentives for National Researchers; and
h. Strengthening the National Agency for Evaluation and Accreditation of Higher Education, and the National System for the Evaluation of the Quality of Education.

Since its establishment in 2012 through 2016, the *FEEI* has invested approximately U$S100 million in approved projects, many of them oriented to teacher preparation and development (FEEI, 2017).

The first program to be launched, and with U$S125 million, was the Paraguayan Program for the Development of Science and Technology (*PROCIENCIA*) implemented by *CONACYT.* Its goal is to strengthen the national capacity in scientific research and technological development. This program has promoted research to an extent, never before experienced in the country. Research and graduate study are being funded. Two projects in Mathematics Education can be highlighted. One is the already mentioned Masters in Science Teaching with specializations in Mathematics Teaching, Physics Teaching and Chemistry Teaching at the National University of Concepción. The other is the Multidisciplinary Organization for Support to Teachers and Students (OMAPA) which has been funded for an in-service preparation of elementary teachers and for mathematical problem solving in Guaraní.

Among the most important results of *PROCIENCIA* since its creation in 2016 are the following:

- ***PRONII***, which already existed, was put under *PROCIENCIA* and thanks to the funds that were received, the 238 researchers that were registered in 2011 grew to 516 in 2016: 248 men and 268 women. These researchers are principally responsible for moving forward the development of the other *PROCIENCIA* projects;
- **Project I + D**, has reviewed 512 research proposals and 323 of them are being implemented;
- **Graduate Programs**, 48 graduate programs have been presented by public and private universities from throughout the country and 19 of them are functioning with 337 participants;
- 77 scientists and technologists have received travel funds to travel outside the country to learn more about scientific research and research management;
- The call for the creation of **Offices for Technology Transfer and Research Results** with 10 proposals under review; and
- The **Center for Scientific Information** at *CONACYT*, with the creation of a web portal that makes high level scientific information accessible to the scientific and education community.

In 2015 five important public programs were approved: (1) early childhood care from pregnancy to five years old, (2) textbooks for upper secondary schools, (3) educator development (particularly with respect to integrating technology throughout the education system), (4) financing the "Don Carlos Antonio López" Scholarship (*BECAL*), and (5) increasing investment in research and science (as well as making sure that such funds are available as intended).

Among the projects that have been funded, the most ambitious with respect to teacher specialization is called Development of Educators for Improving the Learning Boys, Girls, Youths and Adults at the National Level (FEE, 2014). Its goal is to strengthen teacher classroom competence through professional development. The program is for both elementary and secondary teachers. The investment approaches U$S49 million. Despite its importance, its implementation did not begin until December of 2016. It should last for four and a half years. It involves Ministry personnel, national and international universities, teacher preparation institutes, and other organizations, as well as national and international academics. There is no precedent in the country for an in-service teacher development program on this scale.

Also, the *BECAL* program is concerned with professional development in various forms. Since its implementation until the first quarter of 2017, 844 scholarships had been awarded: 122 for a Masters in Education, 100 sent public school teachers to Spain in 2016, and another 80 to France in 2017. Managers at *BECAL* and the Ministry are articulating efforts to create the necessary conditions for an adequate transfer of knowledge when the scholarship recipients return to their positions (BECAL, 2017).

The project for integrating technology throughout the education system is receiving an investment of US$127 million. It includes not only provision of equipment and facilities, but also the necessary training for teachers, administrators and parents in the use of those resources. It is hoped that this project will be implemented during the second half of 2017.

The annual audit of FEEI in 2016 showed that 13 public programs with an allocated investment of U$S514 million had been approved. Although as of this writing only 25% of the funds have been transferred to the recipients, the balance is committed and protected for the exclusive use of these programs, some of which will be in operation until 2021.

Five years after the passing of the *FONACIDE* and *FEEI* laws the use of the funds for improving teacher preparation is just being initiated.

References

Banco Interamericano de Desarrollo (BID). (2012). El camino hacia el éxito en matemáticas y ciencias: Desafíos y triunfos en Paraguay. Nuevos hallazgos del Banco Interamericano de Desarrollo sobre el pensamiento crítico en la educación pre-primaria y primaria. Washington, DC.: Banco Interamericano de Desarrollo-División de Educación. Recuperado de http://tiny.cc/135ocy.

Banco Mundial. (2013). Estado del arte de la profesión docente en Paraguay. Ideas inspiradoras para la elaboración de políticas educativas. Asunción: Autor. Recuperado de http://tiny.cc/aw5ocy.

Becas Don Carlos Antonio López (BECAL). (20 de febrero de 2017). Becal Entrega 80 contratos a docentes—"Viva el Paraguay". Autor. Recuperado de http://www.becal.gov.py/becal-entrega-80-contratos-a-docentes-viva-el-paraguay/.

Congreso de la Nación Paraguaya (CNP). (26 de mayo de 1998). Ley General de Educación. [Ley 1264 de 1998].

Congreso de la Nación Paraguaya (CNP). (07 de junio de 2001). Ley que reconoce al Instituto Superior de Educación "Dr. Raúl Peña" de la Ciudad de Asunción, como una Institución de Educación Superior. [Ley 1692 de 2001].

Congreso de la Nación Paraguaya (CNP). (2012). Inversión Pública y Desarrollo (FONACIDE) y el Fondo para la Excelencia de la Educación y la Investigación. [Ley 4758 de 2012]. Recuperado de http://www.hacienda.gov.py/fonacide/archivos/documentos/Ley4758-12.pdf.

Congreso de la Nación Paraguaya (CNP). (02 de agosto de 2013). Ley De Educación Superior. [Ley 4995 de 2013].

Durán, M. (2012). La educación en el Paraguay colonial. Asunción: ABC COLOR. Recuperado de http://tiny.cc/ix5ocy.

Falus, L & Goldberg, M. (2011). Perfil de los docentes en América Latina. In CUADERNO 09 de SITAL. Buenos Acres: IIPE-UNESCO-DEI

Fondo para la Excelencia de la Educación y la Investigación (FEEI). (28 de octubre de 2014). Modificación del Programa Capacitación de los Educadores para Mejoramiento de los Aprendizajes de Niños, Niñas, Jóvenes y Adultos a nivel Nacional. Autor. Recuperado de http://fondodeexcelencia.stp.gov.py/aprobados/vuip-f-77.

Fondo para la Excelencia de la Educación y la Investigación (FEEI) (2017). Informe de Gestión y Avances 2016. Autor. Recuperado de http://www.fondodeexcelencia.gov.py/wp-content/uploads/2016/12/INFORME-2016-FEEI-2-1.pdf.

Martínez, M. (noviembre, 2013). Reflexiones sobre el profesorado. La formación docente, una realidad postergada. Revista Paraguaya de Educación, 1(3), 2013, pp. 17–28. Recuperado de http://tiny.cc/sz5ocy.

Ministerio de Educación y Cultura (MEC). (2000). Resolución 10747/00. Suspensión de la apertura y habilitación de nuevos institutos de formación docente de gestión oficial y privada.

Ministerio de Educación y Cultura (MEC). (2009). Estructura del Sistema Educativo Nacional. Recuperado de http://tiny.cc/6y5ocy.

Ministerio de Educación y Cultura (MEC). (2011). Matemática. Módulo 1: Trigonometría. Campaña de Apoyo a la Gestión Pedagógica a Docentes y Directivos en Servicio. Asunción, Paraguay: Autor.

Ministerio de Educación y Cultura (MEC)—Viceministerio de Educación Superior. (Abril 2012—2da edición). Datos sobre la Educación Superior en el Paraguay. Asunción, MEC.

Ministerio de Educación y Cultura (MEC), Juntos por la Educación. (2013). Mesa Temática. Formación Docente 2012–2013. Asunción, Paraguay: Autor.

Ministerio de Educación y Cultura (MEC). (2013). Diseño Curricular. Profesorado de Educación Escolar Básica Primero y Segundo Ciclos. Asunción, Paraguay: Autor.

Ministerio de Educación y Ciencia (MEC). (2017) Portal oficial. Sistema educativo. Recuperado en http://www.mec.gov.py/cms/?ref=297778-calendario-educativo-nacional-2017=296669-calendario-escolar-2016.

Organización de las Naciones Unidas para la Educación, la Ciencia y la Cultura (UNESCO). (2012). Antecedentes y Criterios para la Elaboración de Políticas Docentes en América Latina y el Caribe. Santiago de Chile. Autor. Recuperado de http://www.minerd.gob.do/idec/Docs2/6d-CriteriosPolDocentesALCINFOFINALUNESCO18512FINAL(3).pdf.

Organización de las Naciones Unidas para la Educación y la Cultura (UNESCO). (1985) Educación en el Paraguay. Principales Indicadores Seguimiento Periodo: 1981–1983. Recuperado de http://www.ibe.unesco.org/National_Reports/Paraguay/nr_mf_py_1984_s.pdf.

Oficina Regional de Educación para América Latina y el Caribe, Organización de las Naciones Unidas para la Educación, la Ciencia y la Cultura (UNESCO). (2014). Primera entrega de resultados del tercer estudio regional comparativo y explicativo. Comparación de resultados del segundo y tercer estudio regional comparativo y explicativo 2006–2013. Santiago de Chile. Recuperado de http://tiny.cc/105ocy.

Organización de Estados Iberoamericanos, Ministerio de Educación y Cultura [OEI-MEC]. (1994a). Breve evolución histórica del sistema educativo. Sistemas Educativos Nacionales. Paraguay (Capítulo 2). Recuperado de http://tiny.cc/c15ocy.

Organización de Estados Iberoamericanos, Ministerio de Educación y Cultura [OEI-MEC]. (1994b). La educación superior no universitaria – Formación docente. En Sistemas Educativos Nacionales. Paraguay (Capítulo 11). Autor. Recuperado de http://tiny.cc/y15ocy.

Poder Ejecutivo de Paraguay (PEP). (16 de enero de 1968). Creación del Instituto de Educación Superior Dr. Raul Peña. [Decreto 31.003 de 1968].
Poder Ejecutivo del Paraguay. (2 de octubre de 2003). Decreto por el cual se reglamenta la aplicación de la ley nº 1725, del 13 de setiembre de 2001, del estatuto del educador. [Decreto Nº 468 de 2003].
Reimers, F. (1993). *Análisis del Sistema Educativo en el Paraguay: Sugerencias de política y estrategias para su reforma*. Asunción del Paraguay: Harvard Institute for International Development/Centro Paraguayo de Estudios Sociológicos.
Rivarola, D. (2006). Paraguay: Reforma Educativa y Crisis de la Docencia. Asunción: Organización de Estados Iberoamericanos. Recuperado de http://tiny.cc/j25ocy.
Rojas, V. S. (2011). *La condición docente de la educación inicial y escolar básica*. Conec: Recopilación de datos y sistematización. Asunción.
World Economic Forum. (2013). The Global Competitiveness Report 2013–2014: Full Data Edition. Recuperado de http://tiny.cc/h35ocy.
Vaillant, D. (2009). *"Políticas para el desarrollo profesional docente efectivo", en Aprendizaje y desarrollo profesional docente*. España: OEI-Fundación Santillana.
Zayas Rossi, L. I. (2015). Historia de la formación docente en Paraguay. *Praxis educativa, 19*(3), 32–44.

Report on Mathematics Teacher Preparation in Perú

Augusta Osorio Gonzales, María Aredo Alvarado, María del Carmen Bonilla Tumialan, Olimpia Castro Mora, Lilian Isidro Cámac, Cerapio Quintanilla Cóndor, Carlos Sabino Escobar and Martha Rosa Villavicencio Ubillús

A. Osorio Gonzales (✉)
Pontificia Universidad Católica del Perú, Lima, Peru
e-mail: arosorio@pucp.edu.pe

M. A. Alvarado
Universidad Nacional de Piura, Castilla, Peru
e-mail: maria_aredo@hotmail.com

M. C. Bonilla Tumialan
Asociación Peruana de Investigación En Educación Matemática (APINEMA), Coordinadora de La Red Latinoamericana de Etnomatemática-RELAET- Capítulo Perú, Lima, Peru
e-mail: mc_bonilla@hotmail.com

O. Castro Mora
Ministerio de Educacion, Lima, Peru
e-mail: olimpiacastro@hotmail.com

L. Isidro Cámac
Especialista de Currículo Del Ministerio de Educación, Lima, Peru
e-mail: lilian_isidro@hotmail.com

C. Quintanilla Cóndor
Universidad Nacional de Huancavelica, Huancavelica, Peru
e-mail: quintanilla_cn@hotmail.com

C. Sabino Escobar
Instituto de Investigación Para La Enseñanza de La Matemática de Tumbes (IREM—Tumbes), Universidad Nacional de Tumbes, Tumbes, Peru
e-mail: csabinoe@gmail.com

M. R. Villavicencio Ubillús
Dirección de Educación Intercultural Bilingüe, Ministerio de Educación de Perú, Lima, Peru
e-mail: villavicenciomr@gmail.com

Y. Y. Baldin and U. Malaspina (eds.), *Mathematics Teacher Education in the Andean Region and Paraguay*, https://doi.org/10.1007/978-3-319-97544-3_4

Introduction

Perú is a sovereign nation located on the west coast of South America, and is bordered by the Pacific Ocean to the west, Ecuador and Colombia to the north, Brazil to the east and Chile to the southeast. Its area of 1,285,215 km^2 is composed of the three regions: the mountains with valleys, plateaus and the high summits of the Andes that comprise 28% of the territory; the coast, a desert region, that is 12% of the territory; and the Amazon Jungle that is 60% of the territory. Perú has among the greatest biological diversity in world and among the greatest mineral resources. The population is over 31 million, 84% of whom speak Spanish, while 13% speak Quechua and 2% Aimara.

The Historical Context of Teacher Preparation

The following summary of the historical development of the preparation of teachers in Perú is supported by a report from the Organization of Ibero-American States (Perú. OEI, 1994).

The Regulations for Instruction of 1850, passed during the first government of President Ramón Castilla, called for the creation of the Central Normal School in the capital of the Republic. The process of creating that school began in 1852, but it was not completed until 1859. It was closed in 1869 without having many graduates despite a large number of students. In 1876, President Manuel Pardo created the Women's Normal School which was located in the old Convent of San Pablo. It was directed by French nuns of the Sacred Heart and functioned as such until 1928 when it became the National Pedagogical Institute for Women. It continues today as the Monterrico National Pedagogical Institute.

On January 28, 1905, during the government of José Pardo, the creation of the Men's Normal School in Lima was decreed to prepare elementary teachers. In 1929, the name was changed to the National Pedagogical Institute for Men and its mission increased to include the preparation of secondary teachers. In 1951, it began to prepare teachers for Technical Education. An agreement with the Peruvian-North American Cooperative Education Service (*SECPANE*) made possible the construction of its current campus in La Canuta, Chosica. On July 6, 1953, it opened as the Central Normal School for Men and changed its name again in 1955 to the Enrique Guzmán y Valle Higher Normal School. In 1965 (and officially on May 23, 1967), it became the National Education University.

At the university level, the preparation of teachers has its antecedents in the professorship in Pedagogy at the University of San Marcos beginning on March 18, 1876. In 1901, the professorship in Pedagogy included a required course in the curriculum of the Faculty of Philosophy and Letters. On December 14, 1925, the Pedagogy section was created in the Faculty of Letters, but it was not until April 24, 1946, that the Faculty of Education was created at the University of San Marcos. The Catholic University, a private university created in 1917, established a

Higher Section of Pedagogy in 1936, which was the nucleus for the formation of the Faculty of Education in 1947. Parallelly, other Faculties of Education were opened in Arequipa, Trujillo and Cusco.

In 1950, the National Plan for Education was approved, creating rural and urban Normal Schools that were distributed in the north, center and south of the country. Some of the Normal Schools were administered by the Ministry of Education and others by religious congregations. However, all of them received federal funds. In the decade of the 1960s, with the proliferation of Normal Schools, there was often an oversupply of teachers. During the military government beginning in 1970 some Normal Schools were closed. Today, the Normal Schools have become Higher Pedagogical Institutes.

With respect to the development of Mathematics Education, Carranza (2007) indicates that in the 1960s the Institute for the Promotion of the Teaching of Mathematics (*Instituto para la Promoción de la Enseñanza de las Matemáticas-IPEM*) was created. The objectives of the Institute were to develop professional development courses for pre-university Mathematics teachers, disseminate teaching ideas and concepts related to the teaching of Mathematics, collaborate in the publication of books, and help national organizations promote the teaching of Mathematics. According to Carranza and Malaspina (2014), the origin of *IPEM* was associated with the Institute of Mathematics of the Engineering University (*Instituto de Matemática de la Universidad de Ingeniería-IMUNI*) which from 1962 to 1968 was dedicated to Mathematics research and leadership development for Perú.

In the 1970s, with the closing of *IMUNI*, a new stage of development in Mathematics began in the Pontifical Catholic University of Perú (*PUCP*). As a consequence, a Masters in the Teaching of Mathematics was created at *PUCP* in the 1980s.

Another milestone in the development of Mathematics Education in the country, in response to changes in the 1970s implied by the teaching of Modern Mathematics, was the team of specialists coordinated by Teresa Arellano Bados in the National Institute for Research and Teacher Development, and the National Institute for Research and Development of Education (*Instituto Nacional de Investigación y Desarrollo de la Educación-INIDE*). Later, the Directorate of Research of what had been *INIDE* contributed in a relevant way in the search for answers to the linguistic and cultural diversity in the Peruvian reality. As the result of studies carried from 1979 to 1988, a proposal for constructing a Mathematics Education with an intercultural and bilingual focus was presented. *INIDE* was phased out in the 1990s during the Fujimori government. Thus, there has been no institution responsible for basic national educational research that is needed for feedback and innovation.

The Structure of the Peruvian Educational System

The Constitution of 1828 indicated that the State was responsible for guaranteeing free primary level instruction to all citizens. Since the first term of Fernando Belaúnde (1963–1968) there has been free public education at all levels.

The General Law of Education N° 28044, in effect since July of 2003, establishes the Ministry of Education as the national government agency that has the final authority to define, direct and articulate policy for education, recreation and sports in accordance with general policies of the State.

The National Education Project (*PEN*) for 2021: "The Education We Want for Perú," approved in January of 2007, proposes six strategic objectives. It is an instrument for the formulation and execution of public policies, and for the mobilization of the citizenry. It serves as the strategic framework for making decisions and as a reference to evaluate the educational action of the State and the society.

The current structure of the Educational System is divided into two stages: Basic Education and Higher Education. Basic Education includes Regular, Alternative, Special, Bilingual Intercultural and Technical Productive. Higher Education is composed of University, Higher Technological, Higher Pedagogical and Higher Schools.

Regular Basic Education (*ERB*) is for children and adolescents who participate in the educational process according to their physical, emotional and cognitive development from the moment of birth. *ERB* is organized into three levels: pre-primary, elementary and secondary. The levels are organized into cycles, where cycles are understood to be educational processes that are developed as a function of advancements in learning. There are seven cycles that begin in pre-primary. Table 1 presents a summary of *ERB*.

Basic Alternative Education (*EBA*) provides education for youth and adults, including adolescents beginning at 14 years of age that need to combine work and study. *EBA* has the same objectives as *ERB* and equivalent quality. It is divided into three levels: beginning, intermediate and advanced. Among the students who participate in *EBA* are those who did not start their formal education on time, did not finish basic education, must combine work and study, desire to continue their education after becoming literate or are too old for Regular Basic Education.

Basic Special Education (*EBE*) has an inclusive focus and assists students who face difficulties in regular school situations. Specialized educational services for

Table 1 Structure of Regular Basic Education (ERB) in Perú

Levels	Cycles	Grades	Ages
Pre-primary	I	0–2 years old	0–2 years old
	II	3–5 years old	3–5 years old
Elementary	III	Grades 1 and 2	6–7 years old
	IV	Grades 3 and 4	8–9 years old
	V	Grades 5 and 6	10–11 years old
Secondary	VI	Grades 7 and 8 (Courses 1 and 2)	12–13 years old
	VII	Grades 9, 10 and 11 (Courses 3, 4 and 5)	14–16 years old

Source Perú. CNE (2010, p. 8)

prevention, detection and opportune attention for children under six years of age are offered, as well as support and advice to inclusive educational institutions.

Bilingual Intercultural Education (*EIB*) is offered at just the pre-primary and elementary levels. Among its goals are to promote the valuing and enriching of the culture and first language of students, as well as Spanish as a second language. It also requires teachers to master the indigenous language where they are teaching as well as Spanish. In 2016 a program was begun as a pilot program to implement *EIB* progressively for indigenous students of *ERB*.

The Ministry of Education made the decision in 2006 to evaluate all the students in Grade 4 who were in schools with Bilingual Intercultural Education. It was intended that the results would be made available to all interested parties so that the results could be analyzed and actions could be planned to improve student learning. The evaluation included tests of reading comprehension in both the indigenous language (Quechua, Aimara, Shipibo-Conibo or Awajún) and Spanish. Although Mathematics has not yet been evaluated, it is expected that it soon will in Grade 4.

Technical-Productive Education is oriented to helping students acquire and develop workforce competencies. It also attempts to promote an innovative culture that responds to the private sector demands. It is intended for individuals who would like to enter or reenter the workforce, and for students of Basic Education. It is organized in a Basic Cycle and an Intermediate Cycle. The cycles are neither sequential or remedial, but terminal. Graduates of either cycle should be ready to enter the workforce.

National Learning Standards

A document called "National Curricular Design" (Diseño Curricular Nacional-*DCN*) was published in 2008. The *DCN* became the guidelines for *EBR*. The approach adopted in the *DCN* can be described as technological based on competencies. Later, with the National Education Project (*PEN*) for 2021 as a framework, various documents like "Progress Maps" (Perú. Ipeba, 2013) and "Learning Progressions" (Perú. Minedu, 2015) were developed. In 2016 a new "National Curriculum" was released. It provides a framework for educational policy for basic education. What students are expected to learn as part of their basic foundation is indicated. The new National Curriculum, reflects the goals and principles of Peruvian education, the objectives of basic education and the National Education Project.

In the *DCN,* Mathematics is presented with three components. For each component there is a detailed description of the contents and capacities to be developed in each cycle. The new National Curriculum uses "Learning Progressions" to describe Mathematics in terms of four mathematical competencies expressed by performances for each cycle and grade. This new proposal is not yet being implemented in all schools, but should be by March of 2018.

On the other hand, an assessment reported by the National Council on Education (Perú. CNE, 2015) points out that there exists a concern about the mathematical learning achieved by students. Each year an evaluation aligned with the normative curricular documents is carried out for Communication and Mathematics in some of the *EBR* grades. These evaluations, known as the "General Assessment of Students" (*Evaluaciones Censales a estudiantes-ECE*) are implemented by the Office for Quality Measurement (*Oficina de Medición de la Calidad-UMC*) of the Ministry of Education. In 2014, tests were taken by 517,000 Grade 2 regular elementary students and Grade 4 students in Bilingual Intercultural Education. In 2015, for the first time, there was an evaluation of students in the second year of secondary education. At that time, for the grade levels considered, a total of 93.8% of the students in 99.7% of elementary schools were tested and 94.4% of students in 99.5% of secondary schools.

The results presented in Table 2 show that only 27% of students in elementary Grade 2 are achieving a satisfactory level in Mathematics. Although the results are quite low, they do represent a 9% increase since 2013. Also, the data show that the achievement gap that used to exist between public and private schools has mostly disappeared. However, the gap for urban and rural schools seems to be increasing. It should be kept in mind that currently many teachers of indigenous children were not prepared to deal with indigenous language and culture. It is hoped that now that teachers are being specifically prepared to work in Bilingual Intercultural Education, that rural performance will improve.

The first year that students in Year 2 of secondary school were assessed with the national evaluation was 2015. That year over 500,000 students in both public and private schools were evaluated to determine achievement levels at the end of Cycle 6 of *ERB* (Table 3).

Less than 10% of Year 2 secondary students scored at the Satisfactory level. A level of "prior to beginning" was introduced and almost 38% of students were in that level. It is of concern that not only public school students, but over 20% of private school students also scored at that level.

The results also show the great gap between in achievement between public and private schools. And the gap for urban and rural schools is even greater in Year 2 of secondary school than in elementary school Grade 2.

Perhaps there should be some reflection on whether or not there should be such testing in more grades so that lack of achievement related to basic concepts can be detected earlier.

The Structures for Initial Preparation of Mathematics Teachers

Article 6 of Law No. 29944 from 2012, referred to as the Teaching Reform Law (*ley de Reforma Magisterial*), indicates that initial teacher preparation can be accomplished either in Higher Education Pedagogical Institutes (*Institutos de Educación Superior Pedagógicos-IESP*) or the schools of education in universities accredited by the

Table 2 Results in Mathematics for Grade 2 for 2014 and 2015

National		ECE 2014			ECE 2015			Differences	
		1[a] (%)	2[b] (%)	3[c] (%)	1[a] (%)	2[b] (%)	3[c] (%)	1[a] (%)	3[c] (%)
		38.7	35.3	25.9	31.0	42.3	26.6	−7.7[d]	0.7
Gender	Male	38.0	33.8	28.2	30.5	42.1	27.3	−7.4[d]	−0.9
	Female	39.5	36.9	23.6	31.5	42.6	25.9	−8.0[d]	2.4
Type of school	Public	39.0	35.2	25.7	30.8	41.8	27.5	−8.3[d]	1.8
	Public urban	32.0	37.9	30.1	24.9	43.8	31.3	−7.1[d]	1.2
	Private	38.0	35.6	26.4	31.6	43.8	24.6	−6.4[d]	−1.8
Area	Urban	33.9	37.2	28.9	27.1	43.8	29.1	−6.8[d]	0.2
	Rural	59.6	27.3	13.1	54.0	33.7	12.3	−5.6[d]	−0.9
Kind of school	"Full grade"[e]	34.1	37.2	28.7	26.9	43.8	29.2	−7.1[d]	0.5
	"Multigrade"[f]	61.5	26.2	12.2	54.9	33.6	11.5	−6.6[d]	−0.8

Source Perú. UMC (2015)
[a]1 is for the Beginning Level of Achievement (*En inicio*)
[b]2 is for the Progressing Level of Achievement (*En proceso*)
[c]3 is for the Satisfactory Level of Achievement (*Satisfactorio*)
[d]The difference is statistically significant at the 5% level
[e]Full grade (*Polidocente*) means that an elementary school has at least one teacher per grade
[f]Multigrade (*Multigrado*) means that an elementary school has more grades than teachers

Table 3 Results in Mathematics for year 2 of secondary for 2015

National		Prior to beginning (%)	Beginning (%)	Progressing (%)	Satisfactory (%)	Average score
		37.6	40.2	12.7	9.5	549
Gender	Male	35.3	40.3	13.4	10.9	554
	Female	40.1	40.1	11.9	8.0	543
Type of school	Public	42.9	40.4	10.5	6.2	537
	Urban public	39.0	42.4	11.6	6.9	543
	Private	22.3	39.6	19.0	19.2	582
Area	Urban	34.0	41.7	13.7	10.5	555
	Rural	65.2	28.5	4.3	2.0	504

Source Perú. UMC (2015)

National System for Accreditation and Evaluation of Educational Quality (*Sistema Nacional de Evaluación y Acreditación de la Calidad Educativa-SINEACE).*

The *IESP* and universities all prepare pre-primary, elementary and secondary teachers. In 2015 there were 262 public and private licensed *IESP*. The most recent available data on universities involved in teacher preparation are from 1999 when there were 38 (Perú, UMC, 2006). To be able to enter a teacher preparation program, prospective students must have completed secondary education and obtain a place in one of the teacher preparation institutes.

Until 2014, the Ministry of Education only oversaw the *IESP*, and the universities depended upon the National Assembly of University Presidents (*Asamblea Nacional de Rectores-ANR*). With the new law concerning universities in late 2014, the *ANR* and its National Council for the Authorization of the Establishment of Universities (*Consejo Nacional para la Autorización de Funcionamiento de Universidades-CONAFU*) were abolished, and the National Superintendence of Higher Education (*Superintendencia Nacional de Educación Superior-SUNEDU*) was established. Attached to the Ministry of Education, *SUNEDU* is responsible for authorizing universities and ensuring the quality of educational services in the universities. Given this new juncture, it is hoped that in the future there will be a more efficient control that will guarantee the quality of the Faculties of Education.

There are two routes to admission to teacher preparation programs. The *IESP* have open competitive admissions processes and other modalities that they establish. The basic norms for the admissions process are set by the Ministry of Education. The *IESP* have a regular entrance process as well as a waiver process. The admissions examination is 30% reading comprehension, 20% mathematical logical reasoning, 35% general knowledge and 15% general culture.

For admissions to university programs the process in most cases includes multiple choice examinations in science and letters or both. There is not a unique national admissions examination that attempts to measure the effectiveness of secondary education. Other routes to university admissions include being top students at a university's pre-university center, or being the first or second highest student in a high school graduating class. Also, some private universities admit students who have been in the upper third in pre-university achievement.

Weaknesses in pre-university education in Perú have led to the creation of what are called "pre-university academies." Training students to pass university admissions examinations is the goal of these academies and other pre-university schools.

The new university law requires that in order to be granted a Bachelor's degree, elementary and secondary teachers must have successfully completed an undergraduate course of studies and a research project, as well as shown competence in a second language, preferably either English or a native language. A Bachelor's degree and a thesis or sufficiently professional work are required to then be granted a Professional title. The Professional title can only be obtained in the university where the Bachelor's degree was granted.

According to regulations in the current Teachers Law (May 2013), only persons with a Professional title can be teachers. Teachers without appropriate teaching titles had until May of 2015 to change their academic situation and thus maintain their jobs.

Elementary teachers teach all subjects that are indicated for that level, including Personal Development, Social Studies and Civics, Communication,[1] Digital Communication, English, Mathematics, Science, Technology, and Art, but not Physical Education and Sports. Secondary teachers, depending upon the university where they graduated could have a major in Mathematics, Mathematics and Physics, or Mathematics and Informatics.

In a study carried out in 2004, it was found that Secondary Education was the second largest of the university degree programs (Díaz, 2007). Results of the National Survey of Homes in 2011, indicate that among the university programs for persons 17 years or older, Education (including Elementary and Secondary) was the most common with 20.6%. However, that does not mean that there is more vocation for Education than for other careers. Díaz and Saavedra (2000) emphasized that one of the most attractive characteristics of teacher preparation programs in Perú is employment stability, especially in the public sector. Thus, although a teaching career does not bring substantial earnings increase over time, with respect to other professions, teaching does offer more employment stability, a lighter workload and more flexible working hours. Teaching is therefore an attractive alternative for high school graduates from an important sector of the population. Also, contributing to the situation is that the *IESP* and the Faculties of Education at the universities do not have a rigorous admissions process and admit a great number of students (Díaz, 2007).

[1] In the case of Bilingual Intercultural Education (*EIB*) for students of Regular Basic Education (*ERB*) who belong to an indigenous group, Communication will be taught in bother the indigenous language and Spanish as a second language. The rest of the subjects will be taught with an intercultural focus.

Thus, it can be concluded that the majority of the high school graduates who have chosen to enter teacher preparation programs do not do so out of a vocation for teaching and have not been among the graduates with the strongest academic backgrounds. Faced with this situation, and as part of policies established by the Peruvian government based on a deep commitment to education and designed to promote access for high achieving youth to teaching careers, beginning in 2014 the "Vocation for Teaching Scholarship" was created. Students who are number one or two in their graduating class or have a grade average of 15 or more out of 20 have access to comprehensive scholarships to study Education in the best universities in the country. In the first call for applications for this scholarship, students who had already begun their higher education studies were permitted to apply. The first scholarship recipients began their studies in March of 2015. For 2017, it became possible to also select to attend accredited higher pedagogical institutes.

The faculty in the *IESP* that teach future teachers are specialists that are either in tenure track or contractual positions. In the case of Faculties of Education in the universities, those who teach future teachers are in departments or faculties according to the structure of each university. Table 4 shows the academic level of the faculty in the *IESP* according to census data from the Ministry of Education in 2015.

According to the new university law, professors teaching undergraduate courses must at least a Master's degree. In the case of the *IESP*, 22% of the faculty should not be teaching.

The document on assessment and recommendations from the National Council on Education (Perú. CNE, 2015), indicates that there is no program for teacher education faculty that permits their professional development and neither do they have access to adequate salaries. The activities that promote development and improvement have not included them as a public priority. Therefore, the differences between in-service teachers and teacher preparation faculty are almost imperceptible.

Table 4 Academic preparation of teacher educators in the various levels at the *IESP*

	Pre-primary	Elementary	Secondary
Completed non-university higher education	48	77	259
Not completed non-university higher education	0	4	16
Completed university undergraduate degree	29	40	234
Not completed university undergraduate degree	11	10	39
Completed master's degree	59	91	403
Not completed master's degree	54	70	375
Completed doctoral degree	8	29	123
Not completed doctoral degree	11	23	85

Source Perú. Minedu Censo a IESP (2015)

The results obtained on the National Evaluation of Graduates of the *IESP* given by the Ministry of Education show that less than 20% of the graduates reach a satisfactory level in the areas that are evaluated (including Mathematics). This has indicated a urgent need to move forward in establishing standards for graduation from teacher preparation programs, as well as generating actions that guarantee that the programs of study contribute effectively to reaching those standards. (Perú. CNE, 2015, p. 33).

In order to assure a quality education in the Initial Preparation of Teachers, the Ministry of Education has developed a plan called "The Plan for Strengthening Public Institutes for Teacher Preparation in 2017." To support that plan Law Nº 30512 (November 2016), "The Law for Higher Education Institutes and Schools, and Public Teaching Careers", was passed. It regulates the creation, licensing, academic programs, management, supervision, and budgeting of the public and private Higher Education Institutes (*IES*) and Higher Education Schools (*EES*). Thus, quality preparation for the integral development of persons who can respond to the needs of the country, the labor markets and the education system will be provided. In addition, the law regulates the development of public school teachers in the public *IES* and *EESI*.

The Content of Initial Preparation

The Directorate of Higher Education Pedagogy develops the Basic National Curricular Design to be implemented in the *IESP*. However, the universities have the academic autonomy to define their own curriculum.

Currently, the *IESP* are implementing diverse curricular designs. They have been modified in light of demands for new approaches and educational changes according to the nature of the programs and specialties that are offered.

The National Basic Curriculum Design (DCBN, 2010) presents the curriculum for each teacher preparation program that is offered at the *IESP*. Elementary teacher preparation and secondary teacher preparation in Mathematics have programs of study that are structured in two stages. The first stage is for four semesters and is oriented to a General Education that permits the future educator to achieve professional competencies related to personal, professional pedagogical, societal and community dimensions as fundamental aspects of an integral preparation. This stage emphasizes disciplinary preparation, and seeks to bring students closer to a knowledge of their environment, as well as to the foundations of the profession. It attempts to develop their observational capacities, reading comprehension, and ability to find and process information with the use of Information and Communication Technologies. In particular, this first level also seeks to provide positive and motivational first contacts with children and the community, and to strengthen the practice of values and, above all, a respect for diversity.

The second stage is from the fifth to the tenth semester. It is focused on specialized preparation, and develops the competencies to manage their field linked to practice with either elementary or secondary students. This stage is one of theoretical analysis and systemization from their lived experience, and of theorization of educational phenomena observed in classrooms. Processes of abstraction and generalization are accentuated. There is an attempt to develop creative, critical and complex thinking, at the same time as a consolidation of autonomous moral judgement and a commitment to education. For those in elementary education a deepening of knowledge in the relevant subjects is sought. In the ninth and tenth semesters there is intensive student teaching and a link to the research that will lead to the degree. In both stages there are seminars and other events that seek to complement the pedagogical, scientific and technological preparation of the students.

Each program is for ten semesters. The semesters are 18 weeks long, 30 h per week, for a total of 540 h. There are 5400 total hours in the program which is equivalent to 220 credits. The first eight semesters consist of face-to-face classes. The last two semesters are a combination of face-to-face classes and student teaching.

As has already been highlighted, all the *IESP* have the same program of studies. Table 5 below indicates the number of Mathematics and Mathematics Teaching courses that are indicated in the Basic National Curriculum Design for elementary preparation as well as for the preparation of secondary Mathematics teachers.

Thus, it can be observed that 25% of the courses in the preparation of secondary Mathematics teachers are Mathematics or Mathematics Teaching. This could be a signal of the weakness of the specialized preparation that they receive. This is in contradiction with the number of hours per week that are assigned to Mathematics by the National Curriculum (2016) in the secondary schools (13.3% of the total) which does recognize the importance of Mathematics for secondary students.

Below the relationship that exists for Mathematics and Mathematics Teaching courses with the total of courses in the teacher preparation programs in some of the Faculties of Education in recognized Peruvian universities is analyzed. As can be observed in Table 6, in the ten semesters of the initial elementary teacher preparation in Faculties of Education in Perú approximately 5% of the courses are Mathematics or Mathematics Teaching. Secondary Mathematics teachers have about 23.8% of their initial preparation courses in Mathematics or Mathematics Teaching (Table 7).

Table 5 Percentage of courses in Mathematics and Mathematics Teaching in the *IESP*

	Mathematics courses	Mathematics Teaching courses	Other courses	Total courses
Elementary Education	4	3	86	93
	4%	3%	93%	100%
Secondary Education with a major in Mathematics	17	5	68	90
	19%	6%	75%	100%

Source Perú. Minedu (2010a, 2010b)

Table 6 Percentage of Mathematics and Mathematics Teaching courses for future elementary teachers in university Faculties of Education

	Mathematics courses	Mathematics Teaching courses	Other courses	Total courses
Universidad Nacional del Centro	2	2	61	65
	3%	3%	94%	100%
Universidad Nacional Enrique Guzmán y Valle	0	2	32	34
	0%	6%	94%	100%
Universidad Marcelino Champagnat	2	2	81	85
	2%	2%	96%	100%
Universidad Privada Antenor Orrego	3	1	66	70
	4%	1%	95%	100%
Universidad Femenina del Sagrado Corazón	2	3	85	90
	2%	3%	95%	100%
Universidad Peruana Cayetano Heredia	1	1	65	67
	1%	1%	98%	100%
Universidad Peruana Unión	1	2	63	70
	1%	3%	96%	100%
Universidad Nacional Mayor de San Marcos	1	3	70	74
	1%	4%	95%	100%
Pontificia Universidad Católica del Perú	1	3	48	52
	2%	6%	92%	100%
Averages	1.8%	3.2%	95%	

It can be seen that the content of the initial preparation of elementary teachers is similar in the *IESP* and university Faculties of Education. Also, the initial preparation of secondary Mathematics teachers is similar in both types of institutions. On the other hand, initial teacher preparation programs face a problem with respect to quality. The existing curricula, approved between 2010 and 2011 for the *IESP*, are outdated with respect to the needs for current teaching-learning processes and are not connected to the curricular reforms that have taken place in pre-university education (Perú. CNE, 2015, p. 32).

The *DCBN* that has been implemented in the *IESP* since 2010 for both future elementary teachers and future secondary Mathematics teachers includes pre-

Table 7 Percentage of Mathematics and Mathematics Teaching courses for future secondary Mathematics teachers in university Faculties of Education

	Mathematics courses	Mathematics Teaching courses	Other courses	Total courses
Universidad Nacional del Centro	19	0	46	65
	30%	0%	70%	100%
Universidad de Piura	12	1	47	60
	20%	2%	78%	100%
Universidad Marcelino Champagnat	15	2	61	78
	20%	2%	78%	100%
Universidad San Ignacio de Loyola	12	1	53	66
	18%	2%	80%	100%
Universidad Nacional Mayor de San Marcos	14	4	56	74
	20%	5%	75%	100%
Averages	21.6%	2.2%	76.2%	

Table 8 Structure of part of the practicums for future secondary Mathematics teachers at the Universidad Nacional de Educación Enrique Guzmán y Valle

	Credits	Hours	Sessions observed	Sessions planned/taught
Practicum I. observation and planning (semester 5)	2	4	8	4
Practicum II. discontinuous practice (semester 6)	3	6	4	4
Practicum III. continuous practice (semester 7)	4	8	4	8
Practicum IV. intensive practice (semester 8)	5	8		12

professional practicums (student teaching). These activities account for about 17% of the program. Additionally, there are hours of observation and other work that the future teachers do in classrooms with students.

As an example, Table 8 below shows how the practicums are provided for future secondary teachers with a major in Mathematics-Informatics at the Universidad Nacional de Educación Enrique Guzmán y Valle.

Additionally, in the ninth semester the future teachers observe and participate in the management and administration of a school, supervised by a teacher. Future elementary teachers have the same experiences.

It should be kept in mind that Table 8 shows an example of the structure of the evaluated practicum hours during the program, but to be evaluated the future teacher

Table 9 Structure of part of the practicums for future teachers at the Instituto Pedagógico Nacional Monterrico

	Observe teacher or peer (h)	Teach the class (h)	Summer practicum (h)	Total in the semester (h)
Semester 5	6	6	30	42
Semester 6	6	6	30	42
Semester 7	30	30	–	60
Semester 8	30	30	–	60
Semester 9	–	160	–	160
Semester 10	–	160	–	160
Totals	72	392	60	524

Table 10 Structure of part of the practicums for future teachers in the Faculty of Education at the Universidad de Huancavelica

	Hours of theory (h)	Hours of practicum (h)	Hours per week (h)	Total (16 weeks) (h)
Observation and planning	1	4	5	80
Practicum II discontinuous	0	5	5	80
Practicum III continuous	0	6	6	96
Practicum IV intensive	0	8	8	128
Totals				384

must have many hours of observation and practice in schools. To have another idea of the total number of practicum hours, Tables 9 and 10 present the situation at two other teacher preparation institutions.

The Continuous Preparation of Teachers

The National Education Project to 2021 (Perú. CNE, 2007) has as strategic objective 3 to have teachers who are prepared for professional teaching. Therefore, there must be an associated Integral System of Teacher Preparation. To implement such a system, clear standards on what is meant by good teaching are needed, as well as an accreditation process for providers of teacher preparation and professional development. Similarly, there exists a need to restructure and strengthen in-service professional development in articulation with initial teacher preparation. An attempt was made to align these objectives with the UNESCO-IBE (2010) document that emphasized that a program of teacher preparation and continuous professional development has as its ultimate goal the organization and development of activities that update, develop and specialize in-service teachers.

Currently there are two directorates in charge of planning for continuous teacher preparation.

- The Directorate General for Teacher Development which is under the Vice Ministry for Pedagogical Management. The Directorate has as one of its functions to design, regulate, guide, monitor and evaluate the system that integrates the policies on teacher evaluation, career pathways, well-being and innovation from a perspective of continuing professional development.
- The Directorate of University Coordination which is under the National Directorate of Higher and Professional Education, and has as one of its functions to promote mechanisms of interaction among the universities and the set of programs in the Ministry of Education.

The National Plan for In-Service Teacher Preparation 2014–2016 (Ley Nº 29944, 2013) established policy guidelines, modalities, national goals and expected results for in-service activities in the country. This plan was approved and is monitored and evaluated by the Ministry of Education in coordination with the Regional Governments.

In-service professional development can be offered by:

- Both regular and technical-productive schools, for their own personnel.
- Accredited institutions of higher education.
- The Ministry of Education.
- Regional Governments.

Programs of continuous preparation have developed using various modalities over the last few decades. They have gone from being face-to-face with an emphasis on teaching methodologies to an expansion of offering by levels, disciplines and language of instruction. These programs have been developed in different modalities including face-to-face, virtual and mixed (Perú. CNE, 2015).

The Ministry of Education has offered various professional development programs including *PLANCAD*, the In-service Preparation Program and *PRONAFCAP* 2007–2011. Table 11 presents details of those programs.

Additionally, the Ministry of Education has more recently implemented two programs:

- *PRONAFCAP* has offered specialization in Communications, Mathematics and Educational Research serving two groups: one in 2012–2014 and another in 2013–2015.
- *PELA*—Program for Training of Trainers of Academic Coaches. From 2010 to 2015 *PELA* had an intercultural bilingual and inclusive focus. In 2014–2015 it collaborated with the Second Specialization in Academic Coaching Program.

In 2014, the Ministry of Education and National Council on Education conducted the National Teachers Survey (*ENDO*) for the first time in 10 years. *ENDO* asked 9698 teachers from both public and private schools about the resources, conditions,

Table 11 Details of in-service professional development programs from 1995 to 2011

Program name	Years active	Hours	Details
PLANCAD	1995–2001	152 per year	Workshops Follow-up classroom visits
In-service Preparation Program	2005–2006	152 per year	Professional development plans based on regional needs
PRONAFCAP	2007–2011	180	Development of communication skills Development of logical mathematical reasoning Mastery of the discipline and curricular content by level

Table 12 Participation in professional development programs offered by the Ministry of Education and others

Program name	Percent of surveyed teachers participating (%)
PLANCAD	34.5
PRONAFCAP	29.2
Ministry of Education courses	42.7
Specializations	15.7
PELA	16.7
Regional and local governments	30.7
None	18.2

Source ENDO (2014)

values and policies that affect their work and professional development. The teachers were also consulted concerning their participation in professional development since 1995 (see Table 12).

It is important to note, as can be observed in Table 12, that 18% of the surveyed teachers had not taken part in any of the professional development activities offered by the Ministry of Education and other institutions concerned with education in the country. Only 43% reported that they had participated in professional development programs of the Ministry of Education even though the goal had been to reach 100%. Also, it was *PLANCAD*, the oldest program, that served the most teachers with the newest programs far from reaching their goal concerning participation.

In Table 13 it can be observed that Strategies and Teaching Methods was the most requested topic with respect to professional development, followed by topics related to teaching materials and assessment.

Table 13 Topics that in-service teachers selected for professional development

Topic for professional development	Percent of surveyed teachers (%)
Disciplinary content	28
Strategies and teaching methods	77
Psychology and culture	39
Classroom environment	22
Teaching languages	12
Assessment	37
Teaching materials	51

Source Ministerio de Educacion (2014)

The Plan for In-Service Teacher Professional Development 2014–2016 pointed out that there was support from UNESCO and UNICEF for the implementation of the plan. This plan proposed professional development for 100% of teachers by 2016. That meant serving 75,881 teachers at the pre-primary level; 161,108 at the elementary level and 161,981 at the secondary level. There would be special attention to teachers in schools with the lowest achievement levels on the General Assessment of Students and a mixed virtual and face-to-face format would be used.

The document on assessment and recommendations from the National Council on Education (Perú. CNE, 2015) emphasizes that over 40,000 teachers have received online professional development on teaching methods. The Pontificia Universidad Católica del Perú (*PUCP*), the Universidad Peruana Cayetano Heredia (*UPCH*) and the Universidad Continental have been in charge of that professional development.

With respect to the teachers of Mathematics, elementary teachers have received continuous professional development in Communication, Mathematics and Civics. Secondary Mathematics have received professional development on the teaching of Mathematics.

A document presented by the Ministry of Education (Perú, 2014) indicated that the programs of continuous professional development over the last 40 years have lacked sustainability and continuity. While high-profile, they were short-term and one-off. They lacked the creativity to overcome the obstacles of Peruvian geography; the far-flung locations of schools; the weak professional preparation and low self-esteem of classroom teachers; and the scarce resources available to combine with local regional and national efforts. They did not have the imagination necessary to adapt the professional development strategies to reality.

Programs of continuous professional development attempted to overcome deficiencies in initial preparation and have yet to be able to become part of a system of continuous development. (Perú. CNE, 2011, p. 27). One of the difficulties detected in this experience is that the demands for continuous professional development exceed the capacity of the institutions that could provide quality professional development.

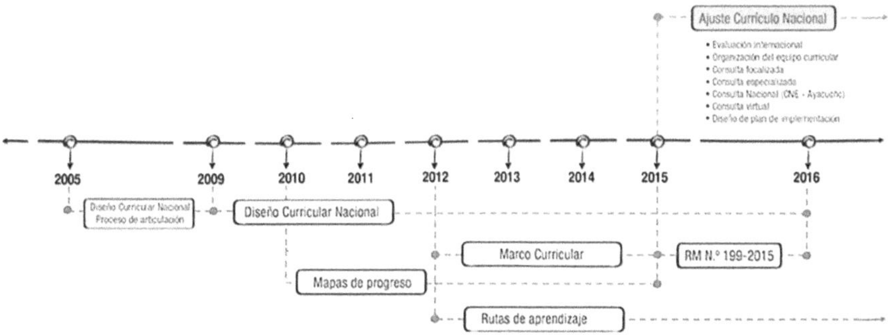

Fig. 1 Changes in curricular design in Perú since 2005

Notable Recent Actions in Initial and Continuing Teacher Preparation

National Curriculum Framework

In recent years, the greatest impact faced by initial and continuing teacher development in Perú has been the coexistence of different curricular proposals from various state organizations.

In Fig. 1 the various guiding documents related to national learning standards that have been introduced and the associated dates are presented. Currently, there are Learning Progressions (Minedu, 2014) and RM N°199-2015 with descriptions of performance indicators by skill, competence and subject area for each pre-university grade level. There will be an implementation of the National Curriculum (2016) beginning in 2018.

The Ministry of Education has introduced almost all the materials related to the Learning Progressions and they have been widely disseminated. Since 2013 an approach to secondary Mathematics based on problem solving, as proposed in the Learning Progressions, is being implemented at the national level. Specialists from the Local Education Management Units (*UGEL*) and Regional Directorates of Education (*DRE*) directed macro-regional workshops on this new approach.

Along these lines, the Mathematics teachers working in Perú who are participating in professional development activities must not lose sight of the competencies that they need to adequately teach their classes. Therefore, they must continue to prepare for a future in accordance with the proposed changes.

Accreditation Standards for Mathematics Teacher Preparation Institutes

In 2006, Law 228740 was passed creating the National System of Evaluation, Accreditation and Certification of Educational Quality (*SINEACE*). Regulations related to *SINEACE* were published in 2007. Also in 2007, the National Education Project (*PEN*) for 2021 "The Education We Want for Perú" was approved with objective 5 that declares the importance for quality higher education to become a determining factor in the country's growth (Minedu, 2007). The Council on Evaluation, Accreditation and Certification of the Quality University Higher Education (*CONEAU*) was created as an organization under *SINEACE* to delineate accreditation process for university programs including Health Sciences, Education and Law. When University Law 30220 went into effect in 2014 the National Superintendence of Higher Education (*SUNEDU*) was created. The law called for the reorganization of *SINEACE* and the creation of a Directive Ad Hoc Council to give continuity to the accreditation process.

As of March of 2017 only six initial teacher preparation programs in secondary Mathematics had been accredited by fulfilling the three dimensions, nine factors and 96 standards each of which had sources of verification that correspond to the first model of accreditation (Table 14).

Careers in Public Teaching

The state has adopted a policy that teachers will be prepared and will work based on criteria defined in the framework of the new Public Teaching Careers (Perú. CNE, 2015). According to the Teaching Reform Law, the Public Teaching Career ladder is structured in eight levels. The time that teachers must spend in each level varies from three to five years, increasing as the higher levels are approached. Therefore, to reach the eighth level a teacher needs at least 30 years of experience. For teachers who work in schools located in areas classified as rural or in border zones, the time needed in the fourth, fifth, sixth, seventh and eighth levels are reduced by a year.

Thus, the Teaching Reform Law, passed in 2012, requires evaluations of teachers to enter the Public Teaching Career and also to rise through the various levels. Thus, the National Program for Teacher Evaluation and Professional Certification was developed, given that evaluations are necessary to enter the new Public Teaching Career ladder and stay on it.

The Ministry of Education has conducted two special evaluations so that in-service educators who were in the old system could enter the Public Teaching Career Ladder. The first was conducted in 2014 and allowed 24,727 teachers to be placed on the new career ladder. A third of them were placed on the fifth and sixth levels, and other two-thirds on the third and fourth. A second special evaluation was conducted for

Table 14 Universities with accredited initial teacher preparation programs in secondary Mathematics

University	Program	Type of institution	Year of accreditation	Location
Universidad Peruana Unión	Secondary: Mathematics and Physics	Private	2015	Lima
Universidad Nacional Daniel Alcides Carrión	Secondary: Mathematics and Physics	National	2016	Cerro de Pasco
Universidad Nacional Herminio Valdizán	Secondary: Mathematics and Physics	National	2016	Huánuco
Universidad Nacional del Centro del Perú	Secondary: Mathematical Sciences and Informatics	National	2017	Junín
Universidad Nacional de Trujillo	Secondary: Mathematical Sciences	National	2017	La Libertad
Universidad Nacional de Huancavelica	Secondary: Mathematics, Computing and Informatics	National	2017	Huancavelica

school administrators. The first regular evaluation for principals and vice principals was conducted at the end of 2014 (Perú. CNE, 2015).

Among the positive consequences of this change is that teachers who wish to be promoted or enter the teaching career, will tend to worry more about their preparation, and that it is consistent with the requirements sought. However, although we still have evaluations that no teacher has passed, hiring was approved given the need to have teachers for the school year. This does not favor the improvement of education.

These evaluations have led higher education institutions to offer programs to assist teaching in their permanent professional development.

Scholarships for Undergraduate Study

During the government of President Humala (2011–2016), specifically in 2014, the National Program of Scholarships and Educational Credit (*PRONABEC*), launched a special program called "Vocation for Teaching." The program is directed at distinguished secondary school graduates who have graduated since 2011 and are interested in a career in Education in any specialty. The students who receive the scholarship can apply to any one of nine private universities that offer Education programs. The scholarship includes not only tuition, but also room and board (when necessary). In the 2016 version of the program 742 students from across the country received the scholarship to begin their studies, and at least 10% chose secondary Mathematics.

Among the positive consequences of this scholarship program, is the hope that within a few years the teachers entering the system who have benefited from the program will be teachers with all the potential to do a magnificent job in the classroom, given the preparation they will have received combined with their interest in teaching.

Additionally, since 2014, Intercultural Bilingual Education scholarships, directed at students from Andean and Amazonian indigenous communities have been offered. Selected recipients can pursue studies in pre-primary and elementary education in the private universities and higher pedagogical institutes that offer such programs.

Scholarships for Graduate Study

In 2014, *PRONABEC* launched a special program of graduate study scholarships directed to public school teachers interested in pursuing a Master's degree in Education. Four important private universities in the country offered the Master's degrees: *PUCP, UPCH, USIL* and the Universidad de Piura. It should be mentioned that three of those universities (*PUCP, USIL* and the Universidad de Piura) offered, among their programs, a Master's degree in Teaching Mathematics. In the case of *PUCP*, 28 teachers were admitted to a Master's in Teaching Mathematics with an emphasis in elementary teaching (12) and in secondary (16). In 2015, 100% of the teachers graduated. At the Universidad de Piura in December of 2015, 76 teachers were admitted to their Master's in Teaching Mathematics: 20 in elementary and 56 in secondary.

One of the reasons for the creation of scholarships and reforms to the teaching career ladder is that students deserve the best education guided by the most qualified teachers. Therefore, the best professionals must become teachers and those teachers should be regularly evaluated so that professional development programs can be organized on the basis of the real needs of teachers.

Unfortunately, because of a change in government, this program was only able to accept applicants once, thus limiting the number of teachers who could benefit from it. On the positive side, the universities that offered the scholarship now have Master's programs designed to meet in-service teacher needs.

Mathematics Education Research and Academic Networks in Relation to Initial and Continuing Teacher Preparation

At the beginning of this document the institutions that were created to work in Mathematics Education, above all in research in the field, were presented. According to Flores and Gaita (2014), after *IPEM* and *INIPM* disappeared in 1970s and *INIDE* in the 1990s, changes in Mathematics Education and in related research continued in the Master's in Mathematics Teaching at *PUCP*. The program of studies in that Master's degree has gone through various changes, framing the different stages of development of Mathematics Education in the country. Initially the program was a set of courses in pure Mathematics, with a level of rigor more or less equivalent to the Master's in Mathematics. There was only one course on research at the end of the program. This corresponded to the vision of the Mathematicians of the time who felt that all that was needed to be a good Mathematics teacher was to know Mathematics.

In 2006 there was a radical reform to program of studies with the adoption of a constructivist position with respect to Mathematics and its teaching. Courses were proposed where the students could initiate research based on the study of theoretical and methodological frameworks specific to Mathematics Education. The program also began to have contact with international researchers and invited them to participate in some of the stages of the development of the renovated program (Flores and Gaita, 2014, p. 86).

Currently, the program of studies for the Master's in Teaching Mathematics has a solid disciplinary foundation that is complemented by carrying out research relevant to Mathematics Education. Thus, this Master's degree program has been able to provide, based on the research carried out by its students and professors, an important support for proposals on teaching and research in this field. Their work is currently being disseminated through the main related international events.

Also, based on the development of Mathematics Education in Perú, two Mathematics Education associations have been created. The Peruvian Society for Mathematics Education (*SOPEMAT*) is an association formed by Mathematics educators from the various levels of the Peruvian education system. *SOPEMAT* provides a space for reflection, and disseminates information on innovation projects and research in the field. In 2012, a group of teachers from various education levels created the Peruvian Association for Research in Mathematics Education (*APINEMA*). *APINEMA* encourages research and innovation through academic events and research projects

It is important to note that in Perú there are two Institutes for Research in the Teaching of Mathematics: *IREM-PUCP* in Lima and *IREM* Tumbes. Both institutes are part of a network of teachers of Mathematics from elementary, secondary and higher education, and work in Mathematics teacher preparation as well as research.

At this time, there are research institutes called "Institutes for Education Sciences" in the Faculties of Education in the following universities: the Universidad Nacional Hermilio Valdizan (Huánuco), the Universidad Nacional San Cristóbal de Huamanga (Ayacucho) and the Universidad Nacional Mayor de San Marcos (Lima).

The Peruvian Network of Universities (*RPU*) is an assembly of universities from the various regions of the country. Table 15 shows information on the 15 universities that belong to the *RPU*, as well as some others, that have a Faculty of Education, a master's, a doctorate and a research institute.

With respect to Master's degrees in Mathematics Education and Mathematics Teaching, it can be observed that five universities in the country have such programs, and three of those have direct connections to related research institutes.

Table 15 Universities in *RPU* with master's and doctoral degrees in Education

University	Region	Undergraduate	Master's	Doctorate	Research Institute
Universidad Nacional de la Amazonia Peruana	Loreto	Education: Elementary, Secondary	Education	Education	
Universidad Nacional de Piura	Piura	Education	Education	Education	
Universidad Nacional Hermilio Valdizan	Huánuco	Education	Mathematics Education	Education Sciences	Institute for Research in Education Sciences
Universidad Nacional del Centro	Junín	Education: Pre-primary, Elementary, Secondary	Mathematics Education	Education Sciences	
Universidad Nacional Daniel Alcides Carrión	Cerro de Pasco	Education: Pre-primary, Elementary, Secondary	Mathematics Education	Education Sciences	
Universidad Nacional San Cristóbal de Huamanga	Ayacucho	Education: Pre-primary, Elementary, Secondary	Education	Education	Institute for Research in Education Sciences
Universidad Nacional de San Antonio Abad del Cuzco	Cuzco	Education	Education	Education	
Pontificia Universidad Católica del Perú	Lima	Education: Pre-primary, Elementary	Mathematics Education	–	Institute for Research on Mathematics Teaching
Universidad Nacional Mayor de San Marcos	Lima	Education: Pre-primary, Elementary, Secondary	Teaching Mathematics in Basic Education	Education	Institute for Research in Education Sciences
Universidad Nacional Federico Villarreal	Lima	Education: Pre-primary, Elementary, Secondary	Education	Education	
Universidad Nacional Enrique Guzmán y Valle	Lima	Education: Pre-primary, Elementary, Secondary	Mathematics Teaching	Education Sciences	

(continued)

Table 15 (continued)

University	Region	Undergraduate	Master's	Doctorate	Research Institute
Universidad Nacional del Callao	Lima	–	Education	Education	
Universidad Nacional José Faustino Sánchez Carrión	Lambayeque	Education: Pre-primary, Elementary, Secondary	Education	Education Sciences	
Universidad Nacional de Huancavelica	Huancavelica	Education: Pre-primary, Elementary, Secondary	Education	Education Sciences	
Universidad Peruana Cayetano Heredia	Lima	Education: Pre-primary, Elementary, Intercultural Bilingual Education	Education	–	

Most of the universities have doctorates in Education or Education Sciences without specific areas of research. Nevertheless, it has been determined that they do have students doing research in Mathematics Education, but there is no systematic information on such studies.

Research Networks and Academic Communities

In 2011, an initiative of the professor/researchers and students in the Master's in Mathematics Teaching at PUCP created a study group on Technologies and Mathematics Education (*TEM*). In this context, there have been activities related to teacher preparation and theoretical reflections, such as the first seminar/workshop with the participation of teachers from both public and private schools in Lima.

In 2012, researchers at *IREM-PUCP* created the research group Didactics of Mathematics (*DIMAT*) that has among its objectives to conduct research on didactical phenomena related to the teaching and learning of mathematical concepts, in both pre-university and university contexts; to contribute to the improvement of teaching and learning Mathematics at all levels; and to articulate research carried out by the group with what is being accomplished with the theses in the Master's in Mathematics Teaching. As can be observed, the creation of such groups as *TEM* and *DIMAT* is fundamental in order to advance research in Mathematics Education.

Mathematics teachers and researchers at the university level participate actively in the meetings of *SINEACE* with the goal of defining and establishing criteria, standards and evaluation processes for educational institutions in Perú. The objective of the meeting is to contribute to curricular changes in the country, particularly in Mathematics.

Likewise, professors from various Faculties of Education participate in work groups or consultations related to policies on initial and continuing teacher preparation that are convened by various government agencies and civic organizations (the National Council on Education, the Educational Forum, among others).

Also, the Faculty of Education at *PUCP* was the first to develop a diploma program for a second specialization in Teaching Mathematics at the elementary level. This diploma program has been preparing professionals since 2007 who are working in the Ministry of Education or in other programs for alternative preparation.

Strengths, Weaknesses, Threats and Challenges

According to Díaz (2015), the strength of teacher preparation is in the existence of institutions that seek to reach an acceptable standard of quality. Such institutions can serve as examples of how there can be change for the better.

Since its creation in 2006, the National System for Accreditation and Evaluation of Educational Quality (*SINEACE*) has been charged with accrediting all the professional programs in universities and institutes throughout the country. A program that is not accredited after three evaluations is closed. Since 2013, *SINEACE* has accredited a group of programs from universities and pedagogical institutes, at the national level, that will be initiating the new stage in initial teacher preparation. But there is still one weakness: At this time, only six university programs have been accredited for Secondary Mathematics teachers.

Díaz (2015) presented an analysis of the principle weaknesses of teacher preparation in our countries which are indicated below:

- Inadequate competence among the professionals graduating from the pedagogical institutes (based on results of the Graduation Evaluation 2013, applied by the Ministry of Education, to students finishing their studies at those institutions).
- The diversity of the curricula. There is the curriculum from the Ministry of Education for the Higher Pedagogical Institutes while each university designs its own given the autonomy granted to them by the law of universities.
- Lack of participation by the teacher preparation institutions in the curricular reforms of Regular Basic Education (*ERB*).
- Student teaching and other practicums that are often just a formality. Only a few of the teacher preparation institutions have their own laboratory schools or agreements with a network of schools where future teachers can apply their theoretical preparation.
- Teacher preparation faculty that resist change and seek to reproduce the pedagogical practices that were taught to them.

- A precarious infrastructure at most of the private higher pedagogical institutes. In 2004, 38% were found to be operating in inadequate locations: houses, commercial buildings and other establishments. The opposite situation exists in the public institutes, where 95% of the facilities have been constructed specially to offer professional preparation.

The reflections of the authors based on their experience as teacher preparation faculty permit the expression of the following weaknesses of teacher preparation in Mathematics at the various levels:

- Inadequate preparation on Mathematics topics.
- Lack of preparation on aspects of teaching Mathematics (or on specialized knowledge).
- Lack of knowledge of some specific topics such as probability, and therefore they are not developed in student teaching or are dealt with very superficially.

Challenges for the Future

The challenges and opportunities that the country faces with respect to teacher preparation are clearly aligned with the main weaknesses. Among the challenges to be met are:

- Find elements that permit a unification of the diverse curricula for teacher preparation in the Higher Education Pedagogical Institutes (*IESP*) and Faculties of Education.
- Increase the amount of time dedicated to the study of Mathematics during teacher preparation.
- Incorporate aspects of Mathematics teaching in teacher preparation.
- Improve the preparation of teachers in topics that either being treated superficially or not at all in the schools (for example, probability).
- Establish norms for improving the infrastructure of private higher pedagogical institutes.
- Provide opportunities for the professional development of trainers of trainers.
- Continue to improve the processes for recruiting students to become teachers.
- Improve the competencies of the graduates of the *IESP*.
- Improve student teaching and other practicum experiences in teacher preparation programs.
- Seek the participation of the teacher preparation institutions in the pre-university curriculum reforms.

References

Carranza, C. (2007). *Historia de la Matemática Peruana.* En: Ciclo de conferencias sobre Matemática y Física Educativa. Lima: Universidad Nacional Mayor de San Marcos.

Carranza, C. A., & Malaspina, U. V. (2014). Perú: A look at the history of mathematics and mathematics education. In *Mathematics and its teaching in the Southern Americas* (pp. 363–380). World Scientific Publishing Co. Pte. Ltd.

Díaz, J. (2007). Educación superior en el Perú: tendencias de la demanda y la oferta. Recuperado enero 2016 de: http://www.grade.org.pe/download/pubs/analisis-2.pdf.

Díaz, H. (2015). Formación docente en el Perú Realidades y tendencias. Fundación Santillana. Recuperado enero 2016 de: http://www.santillana.com.pe/seducacion2015_hugo_diaz.pdf.

Díaz, H., & Saavedra, J. (2000). La carrera del maestro en el Perú. Factores institucionales, incentivos económicos y desempeño. Grupo de Análisis para el Desarrollo (GRADE), Documento de Trabajo, No. 32.

Flores, J. V., & Gaita, R. C. (2014). Situación actual de la Educación Matemática en el Perú. Revista de Matemática, Ensino e Cultura, 9(15), 82–95.

Ley N° 29944. (2013). Ley de Reforma Magisterial. Diario Oficial El Peruano, Lima, Perú, 3 de mayo de 2013.

Perú. Consejo Nacional de Educación (CNE). (2007). Proyecto Educativo Nacional al 2021. La educación que queremos para el Perú. Lima: CNE.

Perú. Consejo Nacional de Educación (CNE). (2010). Propuesta de metas educativas e Indicadores al 2021. Lima: CNE.

Perú. Consejo Nacional de Educación (CNE). (2011). Hacia una propuesta de criterios del buen desempeño docente. Lima: CNE.

Perú. Consejo Nacional de Educación (CNE). (2015). Proyecto Educativo Nacional. Balance y recomendaciones 2014. Lima: CNE.

Perú. Diseño Curricular Básico Nacional para la carrera profesional de profesor de Educación Secundaria en la especialidad de Matemática. (2010). Dirección de Educación Superior Pedagógica.

Perú. Informe OEI—Ministerio de Educación. (1994). Breve evolución histórica del sistema educativo peruano. Recuperado de: http://www.oei.es/quipu/peru/per02.pdf.

Perú. Instituto Peruano de Evaluación, Acreditación y Certificación de la Calidad de la educación Básica (Ipeba). (2013). Mapas de progreso. Lima: SINEACE.

Perú. Ministerio de Educación. (2014). Marco Curricular Nacional, propuesta para el Diálogo – Segunda versión. Lima: Minedu.

Perú. Ministerio de Educación (Minedu). (2015). Rutas para el aprendizaje 2015. Recuperado el 19 junio del 2015, de: http://recursos.perueduca.pe/rutas/.

Perú. Ministerio de Educación, Dirección de Educación Superior Pedagógica. (2010a). Diseño Curricular Básico Nacional para la carrera profesional de Profesor de Educación Primaria. Lima: Minedu.

Perú. Ministerio de Educación, Dirección de Educación Superior Pedagógica. (2010b). Diseño Curricular Básico Nacional para la carrera profesional de Profesor de Educación Secundaria en la especialidad de Matemática. Lima: Minedu.

Perú. Ministerio de Educación, Unidad de Medición de la Calidad (UMC). (2006). Comprendiendo la Escuela desde su realidad cotidiana. Estudio cualitativo en 5 escuelas estatales de Lima. Lima: Minedu.

Perú. Ministerio de Educación, Unidad de Medición de la Calidad (UMC). (2015). Evaluación Censal de Estudiantes 2015 (ECE 2015). Lima: Minedu.

Perú. Ministerio de Educación. Viceministerio de Gestión Pedagógica. (2015). Formación inicial docente. Recuperado de: http://www2.minedu.gob.pe/digesutp/formacioninicial/?p=1264.

Perú. Ministerio de Educación. (2007). Proyecto Educativo Nacional al 2021: La Educación que queremos para el Perú. Lima: Minedu

Perú. Ministerio de Educación. (2014). Encuesta Nacional de Docentes de Instituciones Públicas y Privadas. Lima: Minedu

Perú. Sistema Integrado de gestión para la Educación Superior Pedagógica para IESP, ISE y ESFA. (2015). Instituciones de Educación Superior Pedagógicos 2015. Recuperado noviembre 2015 de: http://siges-pedagogicos.pe/instituciones/iesp-ise-esfa-2015/.
UNESCO-IBE. (2010). World data on Education. Données mondiales de l'éducation. Datos mundiales de Educación. VII Ed. 2010/2011. Recuperado enero 2016 de: http://www.ibe.unesco.org/fileadmin/user_upload/Publications/WDE/2010/pdf-versions/Kyrgyzstan.pdf.

Correction to: Mathematics Teacher Education in the Andean Region and Paraguay

Correction to:
Y. Y. Baldin and Malaspina (eds.), *Mathematics Teacher Education in the Andean Region and Paraguay*,
https://doi.org/10.1007/978-3-319-97544-3

The original version of the book was published in 2016 with exclusive rights reserved by the Publisher. As of December 2020 it has been changed to an open access publication: © The Editor(s) (if applicable) and The Author(s) 2018.
The copyright of the individual chapters has therefore changed to: © The Author(s) 2018
All chapters in the book are licensed under the terms of the Creative Commons Attribution-NonCommercial-NoDerivatives 4.0 International License.
Any third party material is under the same Creative Commons license as the book unless specified otherwise below.

The updated version of this book can be found at https://doi.org/10.1007/978-3-319-97544-3

Y. Y. Baldin and U. Malaspina (eds.), *Mathematics Teacher Education in the Andean Region and Paraguay*, https://doi.org/10.1007/978-3-319-97544-3_5

Conclusions

The basic structure of the country reports calls for the authors to summarize in the final section—their reflections about Strengths, Weaknesses, Threats, Opportunities and Challenges concerning the theme of Mathematics Teacher Preparation. This particular feature of the country reports is a valuable contribution to the CANP project for it provides the authors an opportunity to critically revise the main issues of Teacher Education in their own countries, offering also possibilities for comparative analyses with neighboring regions and communities worldwide. The understanding of differences as well as the common issues among participants of the CANP project constitute the basis to construct a collaborative network of teachers, teacher educators, mathematicians, mathematics education researchers, ministry of education officers and policy makers to work together in the solution of educational problems. The consolidation of such networks has been the ultimate objective of each CANP project.

The Community of Mathematics Education of South America (*Comunidad de Educación Matemática de América del Sur—CEMAS*) is the network constructed from CANP 5, and since the beginning it has received strategic support from REDUMATE, the network of CANP 2 for Central America and the Caribbean. The comparative study of the issues of Teacher Education that can be shared by the participants of CEMAS and REDUMATE can highlight the efforts needed for the development of mathematics education in Latin American countries. A recent publication by León (2018) is an example of such a comparative analysis based on the country reports.

The implications of the achievements of the CANP projects can be perceived also in the ICMI evaluation document about the CANP projects (Koch, 2017). This document used the surveys collected from the participants of CANP 1 to CANP 5 as part of data analysis to assess the achievements of the CANP projects in sustaining the ICMI policy for further activities in developing countries. Although the research methodology of this document is not based on the country reports, the published reports as chapters of a cohesive book and discussion groups of

Y. Y. Baldin and U. Malaspina (eds.), *Mathematics Teacher Education in the Andean Region and Paraguay*, https://doi.org/10.1007/978-3-319-97544-3

educational stakeholders concerning the continuation of the CANP project contribute to corroborate the suggestions for developing countries.

Each chapter in this book gives a clear description of main issues of Mathematics Teacher Preparation and Continuing Education, stressing the cultural context and political policies that are certainly particular to each country. Interestingly, the common and main challenge faced by each country appears to be the gap between the initial preparation of teachers and the realities in classrooms. Moreover, every chapter suggests that the main cause of this gap is the weakness of the teacher preparation curriculum at all levels, notably the elementary level. In analyzing more deeply the causes of such weakness, different reports point out the lack of research centers in mathematics education in their countries that should focus on the connection of mathematics content in the curriculum of teacher preparation courses to bridge theoretical pedagogical knowledge with practice in real classrooms.

We believe that in each country it is very urgent, as part of its educational policy, to stimulate research in mathematical education. Research centers and institutes should be created in countries where they do not exist. In countries where such center and institutes do exist, they should be adequately supported. The creation and further development of postgraduate programs in mathematics education in universities should be promoted. Permanent programs of scholarships and other stimuli for in-service teacher professional development should be created. Stimulating the synergy of all these measures, and especially between the academy and the respective ministries of education in each country, will contribute to the achievement of significant advances in the improvement of the quality of the initial preparation of mathematics teachers and therefore in the education of students.

Therefore, those countries with universities and the research institutions already offering graduate programs to improve teacher preparation, and those in which there are initiatives supporting teacher professional development programs and curricular reforms that address contemporary challenges must be advanced and supported by the leadership in each network.

As editors of the book, we hope that the book can contribute to improved cooperation among the educational policy makers and teacher educators in the region, and it can serve as a reference in their decision making as well as to ground their research projects.

References

Koch, L. (2017). *Evaluation of the Capacity and Network Project (CANP) of the International Commission on Mathematical Instruction (ICMI) Full Report.* Recovered from https://www.mathunion.org/icmi/activities/developing-countries-support/capacity-networking-project-canp.

León, N. (2018). Encuentros y desencuentros en la formación del docente de Matemática en América Latina. *To be published in Cuadernos de Educación y Formación en Educación Matemática.*

MIX
Papier aus verantwortungsvollen Quellen
Paper from responsible sources
FSC® C105338

If you have any concerns about our products,
you can contact us on
ProductSafety@springernature.com

In case Publisher is established outside the EU,
the EU authorized representative is:
Springer Nature Customer Service Center GmbH
Europaplatz 3, 69115 Heidelberg, Germany

Printed by Libri Plureos GmbH
in Hamburg, Germany